AIRCRAFT ACCIDENTS

AIRCRAFT ACCIDENTS

A Practical Guide for Responders

Jim Anderson

Jeff Hawkins

Robert Gill

THOMSON
DELMAR LEARNING™

Australia Brazil Canada Mexico Singapore Spain United Kingdom United States

Aircraft Accidents: A Practical Guide for Responders
Jim Anderson, Jeff Hawkins, and Robert Gill

Vice President, Technology and Trades ABU:
David Garza

Director of Learning Solutions:
Sandy Clark

Managing Editor:
Larry Main

Acquisitions Editor:
Alison Pase

Product Development Manager
Janet Maker

Senior Product Manager:
Jennifer A. Starr

Marketing Director:
Deborah S. Yarnell

Marketing Manager:
Erin Coffin

Marketing Coordinator:
Patti Garrison

Director of Production:
Patty Stephan

Production Manager:
Stacy Masucci

Content Project Manager:
Jennifer Hanley

Art Director:
B. Casey

Editorial Assistant:
Maria Conto

Printed in the United States of America
1 2 3 4 5 XX 09 08 07

For more information contact
Thomson Delmar Learning
Executive Woods
5 Maxwell Drive, PO Box 8007,
Clifton Park, NY 12065-8007
Or find us on the World Wide Web at
www.delmarlearning.com

Library of Congress Cataloging-in-Publication Data:

Anderson, Jim, 1948-
Aircraft accidents : a practical guide for responders / Jim Anderson, Jeff Hawkins, Robert Gill.
p. cm.
Includes index.
ISBN-13: 978-1-4018-7910-5
ISBN-10: 1-4018-7910-1
1. Airplanes—Fires and fire prevention. 2. Emergency medical services. 3. Aircraft accidents.
I. Hawkins, Jeff, 1961- II. Gill, Robert, 1955- III. Title.

TL553.7.A54 2008
629.136'8—dc22

2007038293

NOTICE TO THE READER

Publisher does not warrant or guarantee any of the products described herein or perform any independent analysis in connection with any of the product information contained herein. Publisher does not assume, and expressly disclaims, any obligation to obtain and include information other than that provided to it by the manufacturer.

The reader is expressly warned to consider and adopt all safety precautions that might be indicated by the activities herein and to avoid all potential hazards. By following the instructions contained herein, the reader willingly assumes all risks in connection with such instructions.

The publisher makes no representation or warranties of any kind, including but not limited to, the warranties of fitness for particular purpose or merchantability, nor are any such representations implied with respect to the material set forth herein, and the publisher takes no responsibility with respect to such material. The publisher shall not be liable for any special, consequential, or exemplary damages resulting, in whole or part, from the readers' use of, or reliance upon, this material.

NO WORDS

How does anyone describe "It"?

The pulse-pounding rush...
You're racing towards a column of thick black smoke...
People are running away from it...
You rush in as quickly as possible, and give little thought...
Other times your survival instinct shouts: "Turn the other way!!
Run to safety!!"...
Chaos....
The sights.... The smells and sounds a lifetime will never erase....
Can you describe the searing heat? How can you say what it feels like?
Your sweat burns you as it turns to steam inside your clothing while
your world takes on a surreal appearance...

There is no way anyone can put into words the post crash site....
Wreckage that once gracefully roamed the skies in a miracle called flight...
Only to meet an unfair demise... Sights and smells unlike
anything you have ever known...
Sometimes you swear you can feel the presence of human spirits...
but who would believe you?

We share this common thread. Cold hands curled around a warm cup of coffee
shared with a total stranger as we shiver in the dampness.... for now he or
she has become part of your "family"....

It doesn't matter what flag we salute, what language we speak,
or the uniform we wear. Those of us who have experienced these things, we know...
And yet no language... No words, no movie, no book.... Can ever express
what we have shared...

This book is dedicated to the brave souls who are members of this
"special family"....
All who have, or will someday share these things....

—Jim Anderson

CONTENTS

PREFACE

INTENT OF THIS BOOK

Throughout the years, we airfield firefighters would train with our friends assigned to structural and wildland fire agencies. These fine people would often ask if we knew of any publications that covered basic familiarization of aircraft, and if there was an established list of resources that would be available to them in the event of a downed aircraft. Many additional questions arose from these requests. How could they control a fire with their assigned firefighting apparatus—which were non-airport fire trucks? What checklists were required for such an event?

As experienced trainers, we realized that there was a need for a curriculum devoted to aircraft hazards at an awareness level, targeting those who do not have either the funding or time for advanced aircraft fire-rescue courses such as those required of airport firefighters. It is for these non-airport emergency responders that this book was developed. Containing basic information on airplane anatomy, practical information on preplanning, and response as well as termination of the incident, this book equips first responders with the tools they need to actively participate in collaborative efforts to respond to a downed aircraft outside the airfield.

HOW TO USE THIS BOOK

This book is logically organized in a sequence that is easy to follow and understand. Written in a practical reader-friendly format, the book begins with background information and basic airplane anatomy, walks the reader through incident response, and closes with termination of the incident. The following is a quick description of the topics covered in each chapter.

Chapter 1, Embracing the Challenge, begins with important background information, including the unique challenges of aircraft firefighting, understanding your local resources, and familiarization with local airports and aircraft. The chapter also explains the role that a first responder plays within a response to an aircraft accident.

Chapter 2, Airplane Anatomy, provides a rundown on the basics of airplane anatomy, without delving into unnecessary details. It covers general aviation aircraft and touches on some specialized aircraft.

Chapter 3, Aircraft Hazards, focuses on safety and the importance of understanding what hazards may be lurking in a downed aircraft. Responders must not only know how to respond, they must also learn how to *safely* respond in order to avoid endangering themselves and others.

Chapter 4, Response Considerations, covers the importance of preplanning. Knowing the terrain, establishing unified command among several agencies, and establishing Standard Operating Guidelines for such a response are all critical factors in the outcome of the incident.

Chapter 5, Response Tactics and Strategies, delves into the actual response situation. How to approach the aircraft, determine the level of impact, overcome challenges on scene, assist airport firefighters, and deal with mass casualty issues are all topics discussed within this chapter.

Chapter 6, Termination of the Incident, covers a critical step that can easily be overlooked, the importance of collecting witness statements, critical incident stress debriefing, and recognizing the signs of posttraumatic stress disorder.

Appendix A, Additional Activities, offers *Case Studies* and *Tactical Table-Top Exercises* to provide students and instructors with further practice of critical skills. This appendix supplements the activities included at the end of each chapter to encourage the application of the knowledge learned in the book.

FEATURES OF THIS BOOK

This book focuses on a practical approach—providing emergency responders with basic information coupled with advice from the experts, so that they may effectively respond to downed aircraft.

- **Case Studies** integrated throughout the book highlight actual downed aircraft events, reinforcing the topics under discussion and emphasizing important lessons learned. Additional case studies with discussion points are also included in *Appendix A* for additional practice of critical skills.
- **Safety** notes pinpoint the dangers associated with downed aircraft response and outline cautionary measures that are necessary to working safely on scene.
- **Notes** reinforce main points in topics under discussion, facilitating review of important information in the chapter.
- **Sample Forms and Checklists** provide a foundation for establishing protocol for emergency response procedures regarding downed aircraft. For those regions with previously existing procedures, it encourages readers to research and understand them thoroughly prior to such an event. Readers may also wish to evaluate existing procedures for areas of improvement.
- **Activities** encourage readers to apply what they have learned in the chapter. Instructors may choose to conduct activities in class for review and application of skills. Additional activities are included in *Appendix A*.

SUPPLEMENT TO THIS BOOK

An *Instructor's Guide on CD-ROM* is available to instructors teaching an aircraft accident response course. The CD provides various tools for covering important concepts in skills in class, as well as for evaluating student comprehension of the content.

- The **Instructor's Guide** includes everything an instructor needs to prepare for the classroom, including:
  - **Lesson Plans** outline important points in the chapter and correlate to the accompanying PowerPoint presentations.
  - **Answers to Questions** provide answers to the questions in the chapters and allow instructors to evaluate student learning.
- **PowerPoint** presentations, correlated to the Lesson Plans in the Instructor's Guide, outline important concepts in each chapter, while photos and graphics visually enhance classroom presentations and reinforce critical points.
- **Test banks** in ExamView format provide a venue for testing student knowledge of content presented in the book. Set up by chapter, each exam is editable, allowing instructors to add, delete, or edit questions based on the particular needs of their students.

An *Online Companion* is also available. For further information on topics, or new technology and procedures, you may visit our Web site at www.firescience.com, and click on the Online Companion to accompany this book. Here you can discover more information on:

- Munitions and Hazardous Materials
- Aircraft Composite Materials
- Commercial Aircraft
- New Tools and Innovations
- Aircraft Construction Features
- US and NATO Military Aircraft

ABOUT THE AUTHORS

Jim Anderson has taught hazardous materials and aviation firefighting classes at colleges in California. With over thirty years of experience in aviation-related firefighting, he has served in many capacities, eventually attaining the rank of Battalion Chief.

Jim currently works for Oscar Larson & Associates as an instructor and Hazardous Materials Team Leader. He is an FAA-certified Airport Firefighter, has previously served as a member of the Aviation Safety Section for NFPA, and is certified as a Fire Instructor. Jim writes magazine articles and has conducted aircraft fire and hazardous materials presentations for conferences hosted by a wide variety of organizations including the NFPA, NATO, states, airports, and assorted organizations. He teaches and lectures widely throughout the United States.

Jeff Hawkins is currently serving as a Fire Chief in the Middle East. He retired from the United States Air Force after nearly twenty-five years holding various positions ranging from Firefighter to Fire Chief. Chief Hawkins holds a bachelor's degree in Fire Service Management from Southern Illinois University at Carbondale and is nationally certified as Fire Officer IV, Fire Instructor III, Fire Inspector III, Airport Firefighter, and Hazardous Materials Technician.

Robert Gill is currently a Fire Chief with the Pioneer Fire Protection District in California. With over twenty years of experience in the fire service, he has proudly served as a Fire Chief at Central Calaveras Fire and Rescue Protection District, Assistant Fire Chief of Operations at McClellan Air Force Base Fire Department, and as a Fire Protection Specialist at the Beale Air Force Base Fire Department. He is a certified California State Fire Officer, Hazardous Materials Specialist, Hazardous Materials Instructor, Public Educator Officer, and Fire Instructor. In 1999 he received the "Hall of Fame Award" from the California State Firefighter's Association and U.S. Safety Engineering Corporation for contributions to his community, his department, and the fire service.

ACKNOWLEDGMENTS

This book provides easily understood information based not only on your authors but also on other greatly knowledgeable people such as Mr. Tom Stemphoski at Tyndall AFB, Florida, who was the curator of the USAF Crash Rescue Manual. A Master Instructor with unique expertise in aircraft egress systems, his contributions also included a multitude of other subject areas related to the entire spectrum of assorted aircraft. Ms. Diane Baker, who worked as a key member of the USAF Composites Mishap Office at Hill AFB, Utah, provided information about aircraft construction materials, especially advanced composite materials, and newer technologies.

Fire Chief Rick Aday of the Civilian Flight Test Center at Mojave Airport in California, along with Mr. Dale Lahey and Dave Armour of Transport Canada, provided a wealth of technical and tactical expertise. Captain Shannon Jipsen, pilot for UPS air operations, was a reliable resource for information from the pilot's perspective, as well as knowledge about air cargo safety information.

Officer Byron Jobe, of the California Highway Patrol, provided us constant feedback from the law enforcement perspective. Stephen Jo, MD, was a valuable resource providing input from his long and productive years as a physician.

Special thanks go to Maria Conto and "our guiding Starr, Jennifer," of Delmar Cengage Learning, who made this book happen.

Our appreciation to those reviewers who provided insight in the development of the manuscript:

Kevin S. Elmore
Indianapolis International Airport Fire Department
Indianapolis, IN

Bill Keller
West Virginia University Fire Service Extension
Morgantown, WV

Mark Lee
Missouri Extension Fire Rescue
Training Institute
Columbia, MO

Jason Loyd
Weatherford College
Weatherford, TX

Joe Teixeira
Kellogg Community College
Battle Creek, MI

The authors also wish to acknowledge these additional special people:

Nicki Anderson, health care coordinator, wife, and mentor, Orangevale, CA
Tina, Brian, Greg, and Tim…"*never* a dull moment."
Connie Gill, educator, and loving wife, Auburn, CA
Heather K. Gill, daughter, Auburn, CA
Barbara Haas, Aircraft Rescue and Fire Fighting Group
Jason Kelley, Plant Manager, Scaled Composites, Inc., Mojave, CA
W. G. Shelton, Branch Chief, Virginia Dept. of Fire Programs, Glenn Allen, VA

These people, and others too numerous to list, provided guidance that enabled us to try to keep this book in a language that can be easily understood by the non-airport firefighter, police, medical personnel, disaster planners, and those who may simply want to learn the basics of airplanes and how aircraft accidents are managed.

INTRODUCTION

THE BIRTH OF A BOOK

Time flies when you enjoy your work! It seems like yesterday I was a young skinny firefighter beginning a great career in firefighting. After a premature retirement because of air base closures, I took a long vacation, staying with some very close friends on their newly acquired land in the beautiful Iowa countryside. It was a crisp autumn afternoon; everybody was in town except me.

As I sat under a tree gazing at the lush green rolling farmland and nearby stands of forests, an uneasy feeling started to "itch my brain." An hour passed as I sat in contemplation, when I happened to glance skyward at a commuter plane that was beginning its descent, bound for the airport in Cedar Rapids, some forty miles distant. A bolt of lightening and a clap of booming thunder that seemed to shake the pastureland, echoing off the adjacent trees and hills, occurred and an inner voice seemed to say, "Your job is not finished!"

I immediately got up from my "thinking spot" and ran into their modest farmhouse. I managed to scrounge a few sheets of notebook paper and started writing notes, beginning with an outline, as many tics began to flow from my pen. That evening I begged the kids for a spare spiral notebook and continued this thought process.

The process continued—on my flight back to SAC, I scrawled notes on airsickness bags and the margins of an "in-flight magazine" after running out of room in my spiral notebook.

Many people have asked me why has it taken so long? As we continued to teach subject matter related to aircraft emergencies and hazardous materials, we listened to the input and desires of what people from many fire departments as well as other response agencies wanted. Our original book that was printed by Butte College (California) almost ten years ago has morphed and changed, based on what this wide assortment of people have communicated to us.

We have worked diligently to ensure that the information that we provide to you in this book is thoroughly researched, the most current information in the industry, and of course, practical in nature. It is our intent that this book will provide you with the tools you need to respond to that most unexpected event—the downed aircraft.

—*Jim Anderson*

EMBRACING THE CHALLENGE

Learning Objectives

Upon completion of this chapter, you should be able to:

- Understand the importance of aircraft rescue firefighting training.
- Understand how all fires and emergencies are unique, yet alike.
- Understand an aircraft accident can happen anytime or anywhere.
- Understand you may be first on scene and, if so, the initial steps you must take to control an event.
- Understand various players' roles during an aircraft accident.
- Understand the different types of aircraft that may frequent local airports.
- Understand applicable regulations for aircraft rescue firefighting operations.

INTRODUCTION

Watching a huge aircraft lift off the ground and climb toward the clouds is an amazing experience. An airplane in flight is a symphony of systems and subsystems orchestrated to accomplish a common goal: to keep the flying machine flying. With this in mind, we discuss general facts about airplanes and flight, as well as specific information about accidents and how to deal with them.

The larger the aircraft, the more systems it requires to make the miracle called "flight" possible. Under normal conditions, these systems function as they are intended to and are benign. Under unusually or critically stressful conditions, however, these systems become a potential danger to **first responders**. First, therefore, we consider the importance of training to protect rescuers during an aircraft emergency event.

first responder
local and nongovernmental police, fire, and emergency personnel who are responsible during the early stages of an incident for the protection and preservation of life, property, evidence, and the environment

THE IMPORTANCE OF AIRCRAFT RESCUE FIREFIGHTING TRAINING

As an emergency responder, you already understand the importance of undergoing regular training to ensure that you keep abreast of the safest, most efficient, and most logical methods of performing your job.

As you know, information about and methodologies for disaster-**response** tactics in the fields of hazardous materials (or *HAZMAT*), EMS, and firefighting—including **aircraft rescue firefighting (ARFF)**—continually evolve in order to effectively deal with the challenges posed by our ever-changing world. New technologies are frequently introduced, and existing aircraft undergo significant modifications during their service lives. So, it is crucial that you receive regular continuing education and training to provide you with the latest information to help you keep pace with these changes. If you do not stay abreast of these advances, your ability to survive while involved in an aircraft accident environment is seriously jeopardized.

response
activities that address the short-term, direct effects of an incident

aircraft rescue firefighting (ARFF)
formerly called crash fire rescue (CFR)

Note: During the course of your work, you may encounter someone who claims to be an "expert" in HAZMAT or aircraft emergencies. There is no single point of knowledge, however, in the critical arenas of firefighting, rescue, EMS, or HAZMAT. The foundation for any skilled responder's success is planning, study, and regular training.

ALL FIRES AND EMERGENCIES ARE ALIKE, YET UNIQUE

Disasters and other emergencies share common challenges, needs, hazards, and tactics, but each type of disaster poses its unique set of problems, requirements, and solutions. Consider the following scenarios and points:

- At the scene of a C-130 aircraft accident (crash), a volunteer firefighter was overheard telling new firefighters, "Well, it's the same as a truck accident—just think of it as a big truck with wings." His statement was not only inaccurate, but it also could have led to dangerous consequences. True, there are common tactical concerns in any vehicle accident, regardless of the type or size of vehicle, such as firefighting, water supply, extrication of victims, and mass casualties. The majority of aircraft accidents, however, contain more collective and diverse hazards in a smaller, concentrated area than any other kind of transportation accident. In order to save lives and keep rescue crews as safe as possible, it is important to know and understand these hazards, as well as how they influence fire suppression and rescue tactics.

  Emergency responders and disaster planners should, at the very least, have a *fundamental knowledge of aircraft construction and hazards,* the specialized **resources** required to mitigate these dangers, and the fundamental procedures for dealing with aviation accidents.

- On July 19, 1989, a DC-10 crash at the Sioux City (Iowa) Airport proved that preplanning and training result in successful rescue operations with many survivors, when the rescue and suppression operation unfolded as practiced.

  Preplanning is simply *thinking ahead*—planning for operations at a specific **incident** or hazardous occurrence. A major aircraft accident (crash) will tax all of your resources. For example, most aircraft crashes result in a large, fuel-fed fire. Fires may require water to extinguish flames, so your preplanning must include providing the same type and amount of firefighting vehicles required to combat a comparable fire, such as that on a gasoline tanker truck. The following issues should be considered when compiling a preplan:

  — *Water or other fire-extinguishing agents.* Will your water supply be adequate? Does your community have a large-capacity water-supply system? How many water tenders are available? Will you need to alter your response plans to call for more water tenders than outlined in your current contingency plans? Should you use tandem pumping or drafting operations from a nearby river, pond, or lake? What are your concerns when entering a commercial aircraft? How does this differ from techniques in entering a **military aircraft**?

  — *Crash site and specialized personnel.* If an aircraft crashes into a large building or residential area, for instance, you will need structural firefighting apparatuses, rescue squads, medical personnel, and additional law enforcement officers. If the building is multistory, you will need aerial ladder apparatus, and trained high-rise rescue crews.

- ***Standard Operating Procedures (SOPs)*** or ***Guidelines (SOGs)***. An aircraft emergency, like a structural fire or automobile accident, requires quick thinking and action by your response personnel: rapid intervention may

resources
personnel, equipment, supplies, and facilities available or potentially available for assignment to incident operations and for which status is maintained

incident
an occurrence or event, natural or human-caused, that requires an emergency response to protect life or property

military aircraft
any kind of airplane owned and operated by military forces

Standard Operating Procedure (SOP)
outline explicit procedures and policies that must be followed during an emergency response or other job related duties, and regulations

Standard Operating Guideline (SOG)
provides guidance and suggested procedures for managing emergency responses or guidelines for organizational operating practices

take the form of fire suppression, rescue, and stabilizing victims trapped in wreckage. The roles played by you and your response crew should be outlined in your agency's emergency SOPs or SOGs.

These contingency plans are valuable tools that provide a "recipe" for the actions required by each participant in an aircraft incident. A SOP or SOG describes the process by which your department or organization handles various types of emergencies, so that response operations are standardized and consistent. (This subject is discussed in detail in Chapter 6.)

- *Mass casualty incidents.* A commercial or military aircraft accident, like a building collapse, severe auto accident, or large fire, may result in a large number of severely injured victims. This situation is usually categorized as a mass casualty incident (MCI). MCIs require plans for specialized equipment and special human resource considerations, such as crew rotation; **mutual aid**; the presence of critical incident stress debriefing teams and chaplains or other spiritual leaders; and so on. Mass casualty incidents are terrible ordeals for rescuers, but being thoroughly prepared is the best method of dealing with them.

mutual aid
assistance rendered by one agency to another agency

- *Hazardous materials.* Any aircraft accident is likely to involve hazardous materials. Dealing with such materials requires specialized equipment and personnel. Just like railroads, trucking, and the maritime industry, aircraft operations must adhere to Department of Transportation (DOT) and National Transportation Safety Board (NTSB) safety regulations when carrying cargo and passengers.

The Ways in Which Aircraft Accidents Are Similar to Other Emergencies

All emergencies share these basic characteristics or phases:

1. A call for help
2. Response to provide assistance
3. Situation assessment
4. Scene control
5. Action
6. Termination
7. Post-emergency review

At times, these phases overlap almost seamlessly, while at other times, they become rigid dividing lines transforming a large-scale event into individual segments.

The first phase occurs once your **agency** receives a call for help. This call alerts you to the emergency and initiates action. In the second phase, you and your crew travel to the accident location. Once on the scene, the third phase occurs as you assess the entire incident. Large-scale events may require you to

agency
an organization or division of government with a specific jurisdiction or function offering a particular kind of assistance

move around the perimeter (if possible) and visualize what is involved. The fourth phase is critical: Gaining control of any incident is paramount to a successful response outcome, regardless of the event's size or complexity. Scene control of a large-scale event may seem unattainable, however. This is why the **National Incident Management System (NIMS)** exists.

National Incident Management System (NIMS)
establishes standardized incident management processes, protocols, and procedures that are applicable for all responders (federal, state, tribal, and local)

local government
a county, municipality, city, town, township, local public authority, school district, special district, intrastate district, council of governments (regardless of whether the council of governments is incorporated as a nonprofit corporation under state law), regional or interstate government entity, or agency or instrumentality of a local government; an Indian tribe or authorized tribal organization; or a rural community, unincorporated town or village, or other public entity

terrorism
act that endangers human life or is potentially destructive to critical community infrastructure or key resources, in violation of the criminal laws of any nation with jurisdiction where such an act occurs

Local government handles most emergencies. When a community's resources are insufficient to respond to an incident, however—such as during a homeland security incident, whether it is caused by **terrorism** or natural disaster—local governments may call on the **private sector** and/or **nongovernmental organizations (NGOs)** for assistance. When this occurs, NIMS is implemented to establish standardized incident management processes, protocols, and procedures that all responders—federal, state, tribal, and local—use to coordinate and conduct response actions. When responders use the same standardized procedures, you all share a common focus—incident management—rather than wasting critical time determining who should be doing what and when. The NIMS enhances national preparedness and readiness in responding to and recovering from incidents, because it enables emergency teams and authorities at all levels to use a common language and set of procedures.

It is important to be familiar with your agency's roles and responsibilities when participating in an incident for which NIMS is in effect. As with all procedures, the NIMS is flexible and changes as new concerns are identified. You are encouraged to remain updated about alterations to the NIMS via the organization's bulletins and Web site.

After the scene has been controlled, the fifth phase begins. The first senior fire officer to arrive at the scene must begin to execute an ongoing plan of action. *Action* is defined as doing something to favorably change the outcome of the event. It may be a single action, such as directing crews to attack a small fire, or a complex, multifaceted action, such as implementing several operations simultaneously. Once the primary action has been completed, wrap-up actions can take place, such as posting law enforcement or similar security personnel to preserve evidence and prevent looting.

The conclusion of each phase is determined by the changing situation. Usually, however, the sixth phase, termination, occurs after rescue, extinguishment, mop-up, and hazardous materials concerns have been dealt with. This phase includes releasing the various response crews and agencies and transferring authority and the responsibility for any unfinished tasks to the appropriate agency, such as cleanup contractors, aircraft **accident investigators**, the coroner, and so on.

Incident termination does not conclude the response process, however. All incidents, whether small or large, require a seventh phase: post-emergency review. This review provides lessons learned, validates operations, and ensures that all the tasks were completed. Ideally, post-emergency review should be done at the scene and with as many of the first responders as possible. If this is not practical, schedule a debriefing session as soon as possible and invite all of the agencies involved. In addition, conduct an "in-house only" review for your agency's personnel.

private sector
organizations and entities that are not part of any governmental structure

nongovernmental organization (NGO)
a nonprofit entity focused on the interests of its members, individuals, or institutions so as to serve a public purpose, not benefit a private entity

accident investigators
may work for a branch of the military or for the National Transportation Safety Board; tasked with piecing together all clues that may help determine the cause of the accident

The Ways in Which Aircraft Accidents Are Unique

In the worse-case scenario, airplanes are flying potential *hazardous materials incidents*. This is because aircraft contain assorted hazardous fluids and gases, such as fuels, liquid or gaseous oxygen, highly pressurized hydraulic fluids, and lubricating oils—and many flammable contents.

This book explores the hazards associated with airplane construction materials in more detail in Chapter 2. Briefly, however, airplanes are built with materials, advanced composite materials, that become toxic when burned in pooled hydrocarbon fuels or subjected to impact forces. In addition, although all airplanes share common dangers, military, private, cargo, and commercial airplanes each have hazards and tactical considerations exclusive to their respective types. Cargo aircraft, for instance, carry a wide assortment of commodities. Passenger planes, obviously, carry many people in addition to large amounts of baggage, freight, and parcels.

Many commercial passenger and cargo aircraft have unknowingly carried hazardous materials as a result of shippers accidentally or deliberately mislabeling the materials or failing to declare when hazardous goods are being transported. Military aircraft may carry explosive weapons, munitions, exotic equipment, or other dangerous cargo. Because the majority of air crashes happen away from an airfield, the first agencies on scene often are from off-airport fire and law enforcement agencies. Thus, it is critical that your agency is aware of these potential hazards when responding to an accident.

ANYTIME, ANYWHERE

The next time you travel by airplane, look closely at the landscape as it passes below you. Note the many airfields (active and abandoned, big and small) dotting the ground. These airfields have runways that may accommodate a passenger jet in distress that is attempting an emergency landing. Consider, too, the large amount of flatlands highways and farm fields where a pilot flying an airplane experiencing problems, such as mechanical failures, in-flight fires, HAZMAT incidents, or terrorist acts, could try to set down in an emergency. The following examples of actual aircraft incidents illustrate how the unexpected can happen:

- A large Boeing 767 lost power to both of its engines. Unable to reach a large airport, the pilots made a successful emergency landing at an abandoned military airfield in Canada—where recreational activities were in progress. As the huge plane landed on an abandoned runway, the people on the ground scurried out of harm's way.
- A twin-engine DC-9 passenger jet made a successful emergency landing on Georgia State Highway 42. The plane flew through a storm that had shown no indications of danger when monitored by the aircrew and

air traffic control
the Federal Aviation Administration (FAA) division that operates control towers at major airports

air traffic control. Unfortunately, the storm system began producing large hailstones. After ingesting numerous hailstones, the plane lost power to both engines. The pilots managed to successfully land "dead stick" (gliding without power) on the rural highway. The aircraft remained intact until its wing struck some trees and an embankment along the road. At this point, the aircraft broke apart and plowed into a service station. Of the 81 passengers, 21 survived. The first on-scene responders, from rural fire and police departments, successfully managed the disaster.

- Also in Georgia, a 30-passenger commuter plane made a successful emergency landing in a farm field. First on the scene were the rural sheriff's department and fire department, which successfully completed fire suppression, rescue, and medical aid.
- The first arriving firefighters to an F-117 stealth fighter crash at Sequoia National Park, in California, were wild-land firefighters from a rural fire agency—not military or airport ARFF crew.
- When an Air Force U-2 aircraft crashed in the middle of downtown Oroville, California, the first-in fire suppression and rescue was provided by a small city fire department.

There is always the chance that a series of errors, equipment failure, bad weather, or just plain old bad luck can result in an aircraft fire or crash within your response area. Many of us work an entire career without ever encountering a major aircraft accident. This even includes people assigned to busy airfield fire departments. All too often, response agencies located far away from an active airfield are the "first-in," or even the *only,* fire department responding to an aircraft crash.

CASE STUDY

general aviation (GA)
all air activities and aircraft that are not associated with scheduled commercial aircraft operations or military aviation

In early 2004, two **general aviation** (private) airplanes collided over a remote area of Southern California. One crashed into a grove of trees, with fatal results. The second aircraft was struck by the first where the windshield meets the roof of the airplane. The impact resulted in the control panel being broken and pushed back, and the roof being ripped off the airplane. The pilot suffered head injuries, yet was able to successfully locate and land on a dirt airstrip. Because he had trained and preplanned for emergency landings, he displayed outstanding flying skills that saved his life.

The first on scene at both crash sites were California Highway Patrol Officers. The officers also were well trained and performed in an efficient manner. Statistically, most aircraft crashes involve general aviation airplanes.

Figure 1-1 shows that preplanning pays off. In this instance, the aircraft lost power and made an emergency landing on a busy street in a Midwestern city.

Figure 1-1 *As a result of frequent training and emergency preplanning, the pilot of this battered airplane skillfully made a "dead stick" landing on a busy street in a Midwestern city. Despite the severe damage to the plane, none of the aircraft's occupants and none of the people on the ground were injured.*

WHO IS FIRST ON SCENE?

Often, the first responders on scene (or *primary responders*) are law enforcement personnel. An example of this is the police officer in Agana, Guam, who was first on scene the night a Boeing 747 crashed short of the airport and broke apart in rough terrain. The officer was on his own for more than 20 minutes, and his only resources were citizen volunteer rescuers who lived in the crash vicinity. Many survivors were trapped inside the wreckage, and their cries for help could be heard in the darkness. The rescue operation was hampered by extremely rough, wet, tropical terrain. The response by airport and non-airport firefighters, ambulance crews, and police was hampered by narrow roads and the traffic jams caused by volunteer rescuers and onlookers.

critical rescue and firefighting access area the primary response area for airport-based ARFF service

According to NFPA data, approximately 80 percent of all major commercial aircraft accidents occur in the **critical rescue and firefighting access area**. This is defined as "the primary response area for airport-based ARFF services."

■ Note
If you arrive at a scene and find onlookers in dangerous areas, relocate them directing them to complete a simple task that relocates them in a safer area.

Approximately 15 percent of these accidents occur in aircraft runway landing-approach areas. In situations such as these, off-airport (community/mutual air fire and emergency response agencies) are most likely to be the primary responders.

VARIOUS ROLES AT AN AIRCRAFT INCIDENT

■ Note
If a transfer of command occurs, ensure a full face-to-face briefing occurs. Then, announce the transfer of command over the radio.

Each responder has a role during an aircraft incident. If you are first on scene, take command. Take a deep breath and begin actions to favorably change the outcome of the situation you are facing: Perform a thorough site assessment. Gain control of the scene. Delegate inbound responders (or even bystanders) to help with any passengers who may be walking around (for **triage** purposes, this term is **ambulatory**). If you arrive at a scene and find onlookers in dangerous areas, relocate them directing them to complete a **simple task** that relocates them in a safer area. Examples of a simple task are directing traffic or establishing an exclusion zone to isolate the area from onlookers until the arrival of law enforcement.

Once inbound responders arrive, you may perform a transfer of command to a qualified responder. If a transfer of command occurs, ensure a full **face-to-face briefing** occurs. Then, announce the transfer of command over the radio. If you retain command, provide clear and concise directions to fellow first responders. (The subject of command is discussed throughout this book.)

Ensure that the scene is preserved as much as possible for the investigation following the incident. Treat the area like a crime scene: Keep unauthorized personnel out. Keep a log of everyone inside the cordon. Record your actions on paper so you can recall what happened when you are interviewed. Taking these steps helps you give an accurate report during debriefing, rather than trying to recreate the scene from memory. They also assist accident investigators in their task of piecing together all clues that may help determine the cause of the accident.

Again, prepare for tragic events *before* they happen by practicing hypothetical scenarios and doing preplanning, perhaps in the form of a tabletop exercise or discussion. This will help ensure your agency's efficient, skilled response when an actual event occurs.

triage
the process of sorting patients for priority treatment based on factors such as severity of injury, likelihood of survival, and available resources

ambulatory
patients who are able to walk

simple task
removes a person from a danger zone and relocates them in a safer area

face-to-face briefing
an in-person, verbal briefing from one person to another

AIRCRAFT THAT FREQUENT YOUR LOCAL AIRPORT

An essential element of ARFF preparedness is knowing what types of aircraft frequent your area. Whether your agency is close to a military airfield, a commercial airport, or a general aviation (private) airport (or all three), learn about the particular aircraft that fly in and out of these airfields. If you don't have a local airport, check with neighboring communities to find out what aircraft fly into and out of their airports. Study the basic configurations and any other hazards associated with the various aircraft that may be using a flight path above your community. Your local or state emergency management office is a good place to begin compiling the necessary information. This knowledge will help you prepare for emergencies and maintain a realistic and efficient Standard Operating Procedure or Guideline for managing not only aircraft accidents, but also other unexpected emergencies or disasters.

Several excellent reference books, courses, and specialized schools are available for teaching aircraft firefighting and rescue for the many types of aircraft that navigate the skies above you. Many of these publications, however, target only the airport firefighter. There is more than one way to learn about aircraft hazards and managing incident scenes at aviation disasters. As with studying hazardous materials, it is a good practice to reference multiple sources of information.

Many small airports depend on the closest rural/volunteer fire department for aircraft rescue firefighting (ARFF) services. Most small airports are concerned mainly with smaller private and corporate aircraft, and possibly with commuter passenger flights. Even if the airport closest to your agency is small, however, it is important for you to obtain general information about medium- and large-frame aircraft, because a military or commercial transport aircraft may divert to a smaller airport in an urgent situation.

Some so-called "experts" believe that plane crashes are not survivable. This mind set perpetuates apathy and promotes weak disaster plans, poor training, and an unprepared response force. In fact, many aircraft crashes have survivors. This is why the efficient emergency responder trains for the proper fire suppression, rescue-extrication, and incident management associated with aviation accidents.

SETTING STANDARDS FOR AIRCRAFT RESCUE FIREFIGHTING OPERATIONS

National Fire Protection Association (NFPA)
an independent, voluntary, and nonprofit organization whose goal is to reduce the loss of lives and property resulting from aircraft emergencies

Aviation Section
a section of the NFPA; engaged in the design and operation of aircraft and airport facilities, and in protecting against and preventing loss or injury as applied to aircraft and airport facilities

Regulations Applicable to Aircraft Rescue Firefighting

Just as responding to hazardous materials incidents must be done according to specific standards (as outlined in NFPA Standard 472), Aircraft Rescue Firefighting also has standards that require specialized skills and training. The **National Fire Protection Association (NFPA)** is an independent, voluntary, and nonprofit association whose goal is to reduce the loss of lives and property from aircraft emergencies. It is a source of research and education for all subjects relating to fire and its prevention. Composed of various committees, including the **Aviation Section** (which, of course, is of particular interest to ARFF personnel), the association develops codes and recommendations for fire safety standards.

Aviation Section members are engaged in the design and operation of aircraft and airport facilities, and in protecting against or preventing loss or injury as applied to aircraft or airport facilities.

The members include design engineers of aircraft, airport facilities, and fire protection equipment; operators of airport facilities; pilots; airport fire service and fire marshal personnel; and others who are interested in the section's objectives. These objectives include:

- encouraging a greater understanding of the many aspects of safety to life and property from fire as applied to aircraft and airport facilities
- creating an environment in which mutual understanding and cooperation among the many professional disciplines may be enhanced
- creating opportunities for leaders to emerge and to attain recognition by their peers through elective office

CFR
an old term which used to mean "Crash Fire Rescue"

The NFPA's mission outlines performance standards and mandatory training requirements for ARFF (formerly called Crash Fire Rescue [**CFR**]) crews in suppression/rescue techniques. It also defines performance standards for airports and heliports, and the agencies that support these operations.

Understanding how the NFPA standards apply to other involved agencies, such as airports, will better prepare you to assist them if called on as a co-responder. Several pertinent NFPA standards are discussed in the following list.

- NFPA 402: *Guide for Aircraft Rescue and Fire Fighting Operations.* NFPA 402 provides information about aircraft rescue firefighting (ARFF) operations and procedures for airport and structural fire departments. In addition to aircraft fire suppression and rescue, many airport fire departments are responsible for fire prevention activities, emergency medical services (EMS), and structural fire protection for buildings and facilities on airport property.
- NFPA 403: *Standard for Aircraft Rescue and Fire Fighting at Airports.* This standard outlines the minimum requirements for ARFF provided at airports. It does not, however, address the requirements for *other* airport fire protection services.
- NFPA 405: *Recommended Practice for the Recurring Proficiency Training of Aircraft Rescue and Firefighting Services.* This standard contains the required performance requirements that an authority with **jurisdiction** over ARFF must meet in order to ensure that ARFF fire departments maintain an effective level of proficiency at airports.
- NFPA 418: *Standard for Heliports.* This standard specifies the minimum requirements for fire protection for heliports and rooftop hangars, but it does not apply to ground-level helicopter hangars. This standard does *not* address regulations and procedures for temporary landing sites and emergency **evacuation** facilities.
- NFPA 422: *Guide for Aircraft Accident/Incident Response Assessment.* Standard 422 provides a guide for investigation team assistance in all aspects of information gathering and in assessing the effectiveness of aircraft accident/incident emergency response services. The 2004 edition has been modified to include a simplified reproducible accident/incident investigation form, which is included in this document.

jurisdiction
a geographic area or realm of authority; may be based on legal responsibilities and authorities

evacuation
organized, phased, and supervised withdrawal, dispersal, or removal of civilians from dangerous or potentially dangerous areas, and their reception and care in safe areas

- NFPA 424: *Guide for Airport/Community Emergency Planning.* This guide describes the elements of an airport/community emergency plan that require consideration before, during, and after an emergency has occurred. The scope of the airport/community emergency plan should include command, communication, and coordination functions for executing the plan. Throughout this document, the airport/community emergency plan is referred to as the "Plan."
- NFPA 1001: *Standard for Fire Fighter Professional Qualifications.* This standard identifies the minimum job performance requirements for career and volunteer firefighters whose duties are primarily structural in nature.
- NFPA 1003: *Standard for Airport Fire Fighter Professional Qualifications.* This standard outlines job performance requirements for response, airport fire suppression, and rescue crews. It also specifies age requirements, minimum educational requirements, and medical requirements in accordance with NFPA 1582.
- The **Federal Aviation Administration**, in *FAR Part 139* and in the ICAO Airport Services Manual, outlines specific knowledge and skills related to the aviation firefighting environment.

Federal Aviation Administration (FAA) branch of the Department of Transportation (DOT) that promotes aviation safety by establishing safety recommendations, rules, and regulations that involve aircraft and the aviation industry

Another good way to learn about aircraft firefighting is to study actual case incidents, which you can find in firefighting journals and other media. When you study case incidents, ask yourself: "How would I handle a similar incident with my resources? Were any mistakes made, or did any shortcomings occur, during the incident? What lessons were learned by the responders?"

SUMMARY

This information in this chapter can provide you with an elementary understanding of aircraft emergencies. By now, you should be able to understand:

- *The importance of aircraft rescue firefighting training.* This training is crucial, because many potential hazards are associated with a crashed or damaged airplane.
- *All fires and emergencies are alike.* Aircraft carry large amounts of fuels (Class B fires). Firefighting tactics for fuel fires are the same whether the fuel is spilled from an airplane or a tanker truck.
- *The unique properties of aircraft accidents.* Airplanes carry more flammable fuel than automobiles or other ground vehicles. Some aircraft are larger than many homes and have specialized systems (such as hydraulics, radar, dangerous cargo, and weapons systems) or may employ dangerous weapons systems (military).
- *How aircraft accidents can happen anytime or anywhere.* Private, commercial, and military aircraft are in the skies at all times of the day, all days of the week, and in almost all kinds of weather. This chapter's case study showed that an aircraft accident, no matter how severe, can occur anytime at anyplace.

- *The initial steps you must take to control the event if you are the first on scene.* The majority of aircraft crashes are first managed by non-airport firefighters, law enforcement, or citizen responders.
- *The various players' roles during an aircraft accident.* These roles often are complex and include preserving evidence and interfacing with assorted agencies' personnel, including those from federal, military, or state agencies.
- *The different aircraft that may frequent your local airport(s).* A military airfield, for example, hosts combat or military cargo aircraft. Many commercial airports are home to a military air reserve or National Guard unit (such as an aerial refueling squadron of tanker planes), in addition to a variety of commercial and general aviation aircraft. Regardless of what types of aircraft frequent your local airspace, remember that a distressed aircraft will land at *any* airport, if necessary.
- *The applicable regulations for ARFF operations that must be met to obtain certification for commercial passenger airports and these airports' firefighters.* Further, when dealing with a downed airplane, your agency may be subject to applicable laws in compliance with the FAA, NTSB, or military regulations.

KEY TERMS

Accident investigators Investigators may work for a branch of the military or for the National Transportation Safety Board. They are tasked with piecing together all clues that may help determine the cause of the accident. Accident Investigators may be members of organizations such as the Airline Pilots Association and may also be FBI, ATF, CIA, or private industry personnel, or contracted, self-employed consultants.

Agency An organization or division of government with a specific jurisdiction or function offering a particular kind of assistance. In incident command structures (ICS), agencies are defined either as *jurisdictional* (having statutory responsibility for incident management) or as *assisting or cooperating* (providing resources or other assistance).

Air traffic control The Federal Aviation Administration (FAA) division that operates control towers at major airports.

Aircraft rescue firefighting (ARFF) Formerly called crash fire rescue (CFR).

Ambulatory A medical term that refers to patients who are able to walk.

Aviation Section This organization is a section of the NFPA. Its members are engaged in the design of aircraft and airport facilities, the operation of aircraft and airport facilities, and in protecting against and preventing loss or injury as applied to aircraft and airport facilities. In addition to defining codes and standards, this section is concerned with developing and promoting understanding of aircraft, airports, and other aviation sectors and their need for fire safety; as well as with promoting mutual understanding and cooperation related to aviation fire safety.

CFR An old term meaning "Crash Fire Rescue." Currently, in the United States, the acronym "CFR" refers to the Code of Federal Regulations.

Critical rescue and firefighting access area The primary response area for airport-based ARFF service.

Evacuation Organized, phased, and supervised withdrawal, dispersal, or removal of civilians from dangerous or potentially dangerous areas, and their reception and care in safe areas.

Face-to-face briefing Refers to an in-person, verbal briefing from one person to another.

Federal Aviation Administration (FAA) This branch of the Department of Transportation (DOT) promotes aviation safety by establishing safety recommendations, rules, and regulations that involve aircraft and the aviation industry.

First responder Local and nongovernmental police, fire, and emergency personnel who are responsible during the early stages of an incident for the protection and preservation of life, property, evidence, and the environment, including emergency response providers as described in the Homeland Security Act of 2002, as well as emergency management, public health, clinical care, public works, and other skilled support personnel (such as equipment operators) who provide immediate support services during prevention, response, and recovery operations. First responders may include personnel from federal, state, local, tribal, or nongovernmental organizations.

General aviation (GA) Refers to all air activities and aircraft that are not associated with scheduled commercial aircraft operations or military aviation.

Incident An occurrence or event, natural or human-caused, that requires an emergency response to protect life or property. Incidents can include major disasters, emergencies, terrorist attacks, terrorist threats, wildland and urban fires, floods, hazardous materials spills, nuclear accidents, aircraft accidents, earthquakes, hurricanes, tornadoes, tropical storms, war-related disasters, public health and medical emergencies, and other occurrences requiring an emergency response.

Jurisdiction A geographic area or realm of authority; it may be based on legal responsibilities and authorities. Jurisdictional authority at an incident can be political or geographic (e.g., city, county, tribal, state, or federal boundary lines), or functional (e.g., law enforcement, NTSB, military, public health).

Local government A county, a municipality, a city, a town, a township, a local public authority, a school district, a special district, an intrastate district, a council of governments (regardless of whether the council of governments is incorporated as a nonprofit corporation under state law), a regional or interstate government entity, or an agency or instrumentality of a local government; an Indian tribe or authorized tribal organization; or a rural community, an unincorporated town or village, or other public entity.

Military aircraft These may be any kind of airplane owned and operated by military forces.

Mutual aid Assistance rendered by one agency to another agency (see *mutual aid agreement*).

National Fire Protection Association (NFPA) This is an independent, voluntary, and nonprofit organization whose goal is to reduce the loss of lives and property resulting from aircraft emergencies. It is a source of research and education for all subject areas as they relate to fire and its prevention. The association develops codes and recommendations for fire safety standards. It is composed of various committees, such as the Aviation Section (which is of particular interest in this book).

National Incident Management System (NIMS) In compliance with federal law, the NIMS establishes standardized incident management processes, protocols, and procedures that are applicable for all responders (federal, state, tribal, and local).

Nongovernmental organization (NGO) A nonprofit entity focused on the interests of its members, individuals, or institutions so as to serve a public purpose, not benefit a private entity. NGOs are not created by a government, but may work cooperatively with governments.

Examples of NGOs include faith-based charity organizations and the American Red Cross.

Private sector Organizations and entities that are not part of any governmental structure. These include for-profit and not-for-profit organizations, formal and informal structures, commerce and industry, private emergency response organizations, and private voluntary organizations.

Resources Personnel, equipment, supplies, and facilities available or potentially available for assignment to incident operations and for which status is maintained. Resources are described by kind and type and may be used in operational support or supervisory capacities at an incident or at an EOC.

Response Activities that address the short-term, direct effects of an incident. Response includes immediate actions to save lives, protect property, and meet basic human needs, as well as the execution of emergency operations plans and of incident mitigation activities designed to limit the loss of life, personal injury, property damage, and other unfavorable outcomes. As indicated by the situation, response activities include applying intelligence and other information to lessen the effects or consequences of an incident; increased security operations; continuing investigations into the nature and source of the threat, ongoing public health and agricultural surveillance testing processes; immunizations, isolation, or quarantine; and specific law enforcement operation aimed at preempting, interdicting, or disrupting illegal activity, and apprehending actual perpetrators and bringing them to justice.

Simple task This is a task that removes a person from a danger zone and relocates them in a safer area. This task may be directing traffic, serving meals to tired rescuers, or a similar assignment.

Standard Operating Guideline (SOG) A SOG provides guidance and *suggested procedures* for managing emergency responses or guidelines for organizational operating practices. It defers to a person's experience, common sense, and good judgment and provides greater flexibility in making tactical decisions than does a Standard Operating Procedure.

Standard Operating Procedure (SOP) A SOP is more binding than a Standard Operating Guideline (SOG). SOPs outline explicit procedures and policies that must be followed during an emergency response or other job related duties, and regulations.

Terrorism Any act that endangers human life or is potentially destructive to critical community infrastructure or key resources, in violation of the criminal laws of any nation with jurisdiction where such an act occurs. These acts are intended to intimidate or coerce a civilian population; to influence the policy of a government by intimidation or coercion; or to affect the conduct of a government by mass destruction, assassination, or kidnapping.

Triage This term means "to sort out." It is used in emergency medicine, especially during mass-casualty incidents and in crowded hospital emergency rooms. It is the process of sorting patients for priority treatment based on factors such as severity of injury, likelihood of survival, and available resources.

REVIEW QUESTIONS

1. Statistically, most aircraft crashes (accidents) involve what type of aircraft?
2. Which emergency response agency is *likely* to be *first on scene* at an airplane crash?
3. Why is it important to know the topography of your response area?
4. What does the acronym ARFF stand for?

5. On July 19, 1989, a DC-10 crash at the Sioux City, Iowa, Airport. Despite the magnitude of the accident and number of casualties, many passengers were successfully rescued and given prompt medical treatment. The success of this rescue operation was mainly due to what factors? [1]
6. Define the aircraft emergency-related term "dead stick."
7. Why is it important for you to know and understand various aircraft hazards?
8. What is NIMS, and how does it relate to your role as an emergency responder?
9. What is the best means of passing on information when transferring command?
10. Why is it important to train with the airport or military ARFF crews in your community?

STUDENT EXERCISES

1. List all of the airports and helipads within your first-in response area. Indicate the type of airfield (e.g., private airstrip, public general aviation airport, commercial airport, or military airfield).
2. List the kinds of aircraft that fly into and out of these airfields on a regular basis.
3. At this stage, can you identify any training needs that will prepare you to manage the complexities of an aviation accident?
4. Based on the case study in this chapter, can you identify anything specific your agency can do to improve its ability to manage an aircraft accident?
5. List any natural barriers, such as rivers, lakes, or steep terrain, in your response area that may hamper your response.
6. List human-made barriers, such as bridges, in your area that may not be capable of accommodating the weight or size of large firefighting apparatus.

BASIC AIRPLANE ANATOMY

Learning Objectives

Upon completion of this chapter, you should be able to:

- Identify basic aircraft anatomy.
- Understand aircraft size and weight categories.
- List different types and examples of
  - general aviation aircraft
  - military aircraft
  - civil aircraft
  - specialized aircraft
- Identify similarities and differences among aircraft.

skin
outer covering of an aircraft, including the fuselage and wings

fuselage
the main body of the aircraft, to which the wings and tail are attached

cabin
passenger compartment in an aircraft

cockpit
compartment where the pilots sit to fly the aircraft

flight deck
pilot's compartment of a large airplane; also referred to as a "cockpit"

occupants
passengers and aircrew on board an aircraft; also referred to as "souls on board" (SOB)

canopy
transparent enclosure over the cockpit on some aircraft

airframe
parts of the aircraft having to do with the flight: the fuselage, boom, nacelles, cowlings, fairings, empennage, airfoil surfaces, landing gear, and so on

tail
portion of an aircraft that consists of vertical and horizontal stabilizers, rudders, and elevators

INTRODUCTION

Most of us have flown in airplanes and seen them flying above us, but we don't think about or understand what enables them to get off the ground and remain aloft. As an emergency responder, however, you must have a general understanding of how aircraft are made, as well as of the differences among various kinds of airplanes and how to recognize specific components of different aircraft. Under the surface of an airplane (called the "aircraft **skin**") lays hidden dangers to emergency responders. So, the more you understand about how an airplane is made and what it is made of, the safer and more effective your efforts in an aircraft emergency operation will be.

BASIC AIRPLANE ANATOMY

Preplanning an emergency incident, as has been said, is the single most important step in preparedness. A vital part of aviation incident preplanning is learning as much as possible about the aircraft servicing your area before you are called to an aircraft-related emergency.

All aircraft have general characteristics that responders should become familiar with to in order to fully understand your response objectives. The **fuselage** is the main body of an aircraft; it carries people and cargo. Fuel tanks are located within the fuselage of many types of aircraft. Most aircraft can accommodate temporary fuel tanks in the fuselage. This may be necessary when delivering the aircraft to an overseas location or for special missions that may require longer flying time, such as specialized surveillance or weather data gathering. Some larger aircraft contain a fuselage fuel tank below the main floor.

The **cabin** is the passenger compartment. The **cockpit** is the location inside the fuselage where the pilot and other aircrew sit. On large-frame aircraft, this area is often called the **flight deck**.

Some aircraft (primarily military fighters, sailplanes, and a few private airplanes) house **occupants** within a **canopy**, located in the fuselage. A canopy is defined as a transparent "bubble-" or "teardrop-"shaped covering that enables the pilot, crewmember, or occupant to see out of the aircraft.

Most aircraft have windows on the side of the fuselage and of the cockpit. These windows are usually made from a Plexiglas-like material and can be cut with relative ease. The forward windshield on all aircraft is much thicker and more difficult to cut, however. On most high-speed aircraft, the forward windshield is especially difficult to cut through because it is even thicker than the windshields of other aircraft. In fact, the windshields of many high-performance aircraft are made with a special optical glass that is extremely difficult to cut through.

Figure 2-1 shows an **airframe**, which functions much like the human skeletal system: Airframes provide aircraft with structural strength and shape, just as bones do in the human body. Airframe components that run from the **tail** to the

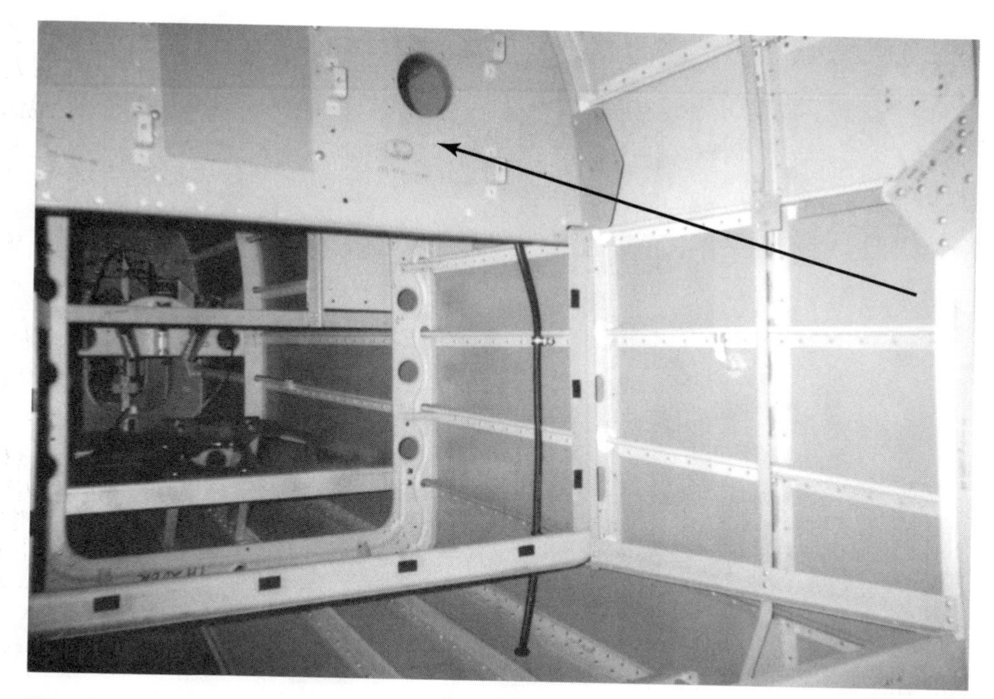

Figure 2-1 *The airframe of a plane works similarly to the bones in our bodies: It provides an aircraft with structural strength and shape.*

longerons
principal longitudinal (lengthwise) structural members of the fuselage

front of an airplane are called **longerons**, sometimes referred to as **stringers**. The figure also shows a **bulkhead**, a circular partition that separates one aircraft compartment from another while giving shape to the aircraft. Understanding the common location of airframe components is important when you are tasked with cutting through the outer skin of a crashed aircraft in order to extricate survivors.

Figure 2-2 shows steps, which may be located on **landing gear** struts, **wing struts**, or exterior fuselage areas to provide access to wing areas, **engines**, or the tail assembly for maintenance or firefighting purposes. These steps provide a means of access to, entry into/exit from, or rescue from an aircraft cabin or cockpit.

• Caution
As an emergency responder, you are more likely to use these steps in adverse conditions such a inclement weather, poor lighting, or slippery conditions from mud, firefighting foam, ice, or snow.

Caution must be exercised when using these steps, because they are relatively small. As an emergency responder, you are more likely to use these steps in adverse conditions such a inclement weather, poor lighting, or slippery conditions from mud, firefighting foam, ice, or snow.

Wheel assemblies of two or more wheels and tires are called landing gear **bogies**, as shown in **Figure 2-3.**

All aircraft have electrical systems, which include batteries, generators, magneto (which generate an electrical spark for the engine cylinders on piston-powered aircraft), electrical wiring and lighting, and an **auxiliary power unit** or **emergency power unit**. Some aircraft may have more than one battery, at different locations in the airplane.

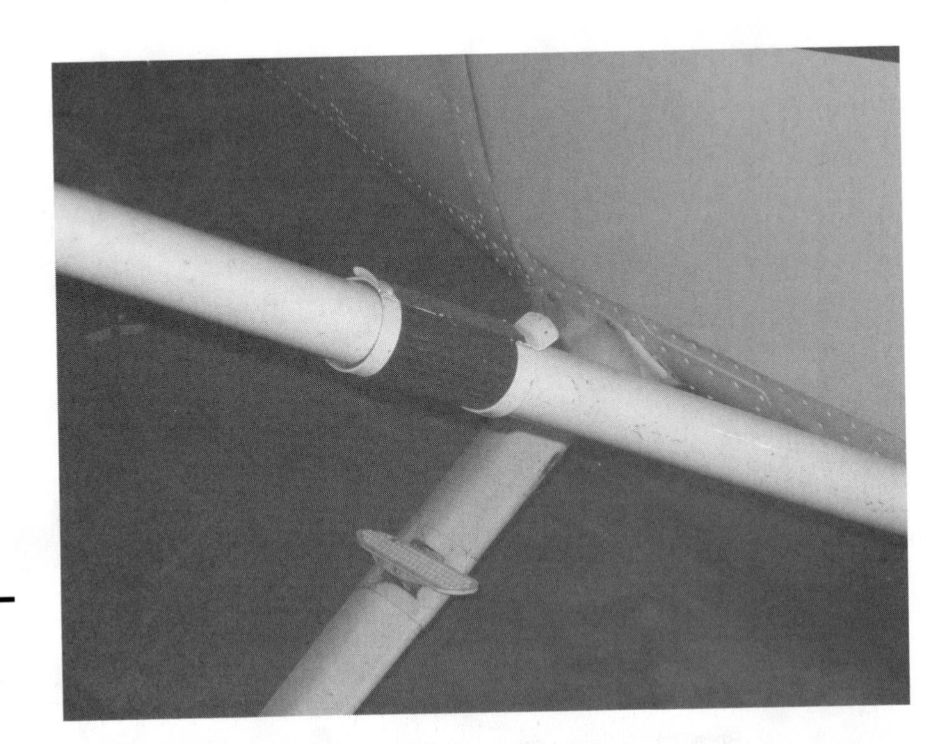

Figure 2-2 *Steps may be located on landing gear struts and wing struts.*

Figure 2-3 *A landing-gear bogie. These are found on high-performance and larger aircraft. A business jet may have two tires on a bogie, while a jumbo jet may have as many as six wheels and tires per bogie assembly.*

Figure 2-4 *Hydraulic lines and reservoirs inside a large passenger jetliner.*

stringer
a long, heavy horizontal timber used for any of several connective or supportive purposes

bulkhead
upright partition that separates one aircraft compartment from another; may carry a part of the structural stress while forming the shape of an aircraft fuselage; equipment and accessories may be mounted on them

Most aircraft have hydraulic systems, which include reservoirs, valves, and tubing. These systems perform tasks such as retracting and lowering the landing gears, moving **control surfaces** to maneuver the airplane, and retracting the tail hooks on military fighter aircraft. Note the fluid lines and hydraulic reservoirs shown in **Figure 2-4.** This portion of the hydraulic system is in the rear fuselage of a **jetliner**.

Figure 2-5 shows a *control panel.* Similar to the dashboard in an automobile, control panels house the gauges and **controls** necessary to operate the airplane. The pilot's control panel is located in the front of the cockpit or flight deck. Older aircraft also may contain flight engineer control panels. Specialized control panels are located elsewhere in aerial refueling aircraft, bombers, antisubmarine, electronic data gathering, and special mission aircraft. Looking at Figure 2-5, you can see that the pilot's hand is on the black knob, which is the engine throttle. To the right is a red knob, which is the fuel mixture control. When this is pulled out (or screwed out counterclockwise), the fuel flow to the engine is decreased.

The engine powers the aircraft. Engines can be designed as piston, jet turbine, or turboprop types. Most single-engine aircraft engines are attached directly

Figure 2-5 *The control panel of a Cessna 172. The pilot is shutting off the engine throttle. This is a "screw-type" throttle, much like the throttle on a fire engine pump panel. Turn the throttle counterclockwise to bring the aircraft engine to idle. The red knob to the right is the fuel mixture control.*

landing gear
the understructure supporting the weight of an aircraft when it is not in the sky

wing strut
looks like a rod or pole and is connected to the airplane between the bottom of the wing and the bottom-side area of the fuselage; provides additional strength for the aircraft

engine
the motive that powers the airplane and allows it to travel

to the fuselage and separated from the cockpit/cabin by a **firewall**, much like in an automobile. The cabin, or firewall, is between the aircraft engine and the cockpit. It is designed to reduce noise and engine heat, and to provide a minimal amount of isolation from an engine fire. The engine on a single-engine amphibious plane is attached to an overhead pylon, which, in turn, is securely fastened to the fuselage.

Figure 2-6 shows a **general aviation aircraft** in the final stages of assembly. Its engine compartment and firewall are clearly visible. This type airplane is equipped with a piston engine, which uses AVGAS, or aviation gas, for fuel. Note that the doors are located in the front and rear of this airplane. This particular airplane also incorporates a movable landing gear, which is hydraulically retracted and extended.

Unlike the piston-powered aircraft in Figure 2-6, some airplanes are powered by turbine engines. Airplanes with turbine engines use Jet A or Jet B aviation fuel. A *turbine engine* is either a jet or a turboprop, which is a turbine powering a propeller, as shown in the general aviation aircraft illustrated in **Figure 2-7.**

Aircraft are designed with entry/exit doors that may be hinged into a "plug-style" configuration, as shown in the figure. This type of door is opened by being pushed from the outside or pulled from inside. Once this is engaged, the door can

Figure 2-6 *A general aviation aircraft in the final stages of assembly. The firewall is clearly visible behind the aircraft engine compartment.*

Figure 2-7 *This twin-engine general aviation aircraft is powered by turboprop engines.*

bogie
a tandem arrangement of landing gear wheels; swivel up and down to enable all wheels to follow the ground as the altitude of the aircraft changes or as ground surface changes

swing open. Side-hinged doors may have an "in-latch" assembly rather than a plug-style assembly. Some aircraft doors are hinged from the top; when opened, these doors drop down. Another door design is called a "clam shell" and is a two-piece unit hinged at the top and bottom. When opened, the top portion swings up, while the bottom portion drops down and has built in steps. Some aircraft, such as the Airbus A-300, are designed with a sliding entry door. Some aircraft, such as the Boeing 737 and MD-80 jetliners, are designed with built-in stairs.

Most aircraft, especially **airliners**, are equipped with inflatable emergency-escape slides, which provide a quick means of exit for passengers and aircrew. These inflatable slides are connected to doors and sometimes to over-wing exits on large passenger airliners. **Figure 2-8** shows how escape slides are typically

Figure 2-8 *The entry door of a typical commercial jetliner. Note the escape slide attached along the bottom edge of the door and the optional built-in boarding stairs.*

● Caution
Stand beside the door when opening it. Some aircraft doors swing open and will strike a rescuer if panicked passengers or temporarily incapacitated aircrew open the door from the inside.

attached to a larger jetliner exit door. Before the aircraft takes off, the cabin crew ensures that it is "armed": They attach a "girt bar" to the bottom of the doorframe. If the door is opened in an **emergency evacuation** situation, this bar stays in place as an anchor, and the slide inflates and is deployed. Figure 2-8 also illustrates how most airlines indicate that an armed slide is attached by using a red strap that is snapped diagonally across the door window and visible from the outside. In some **military aircraft**, such as the KC-10 (a military variant of the DC-10), a red flag is affixed from the inside door or window where any armed slide is located.

Rescue personnel who are opening airline main entry doors from the outside need to exercise caution. Stand *beside* the door when opening it. Some aircraft doors swing open and will strike a rescuer if panicked passengers or temporarily incapacitated aircrew open the door from the inside.

Some emergency escape slides are designed to be used as emergency life rafts for water-landing survival. The locations of the main exit/entry/doors and emergency escape **hatches** over the wings of the airplane are clearly visible. Figure 2-8 also shows built-in stairs, which are available for many aircraft designs, such as this Boeing 737.

auxiliary power unit (APU)
fuel-powered (usually turbine) unit that supplies electrical power, air conditioning, and backup power to an aircraft during flight; may also be used to power pneumatic (air) and hydraulic (fluid) pumps within the airplane

emergency power unit (EPU)
may be used instead of an auxiliary power unit; powered by a toxic fuel called hydrazine; in the event of engine failure, automatically starts and furnishes power for flight instruments and aircraft control movements

control surfaces
ailerons, flaps, elevator, rudder, and spoilers, which control an aircraft's direction of flight, altitude, and pitch

Figure 2-9 shows a rescuer who has laddered the aircraft to the side of the entry/exit door, in order to prevent being injured if the door is opened from inside or if an emergency slide deploys.

Emergency exits include all normal entry/exit doors, service doors, and escape hatches through which passengers and crew can quickly exit (egress) an aircraft. Depending on the type of airplane, the egress system may include one or more emergency-escape slides. Hatches resemble small doors and provide a means for escape from, or emergency entry into, a distressed aircraft.

Figure 2-10 shows a variety of emergency hatches found in private and **commercial aircraft**. They are usually "plug," or "pin-latch," style (see the figures), although a few aircraft incorporate hatches hinged at the top; these are designed to swing out and up. Escape hatches may be located at the sides, bottom, or top of the fuselage. Hatches are usually built into larger general aviation aircraft and commercial and military aircraft that carry passengers or cargo. Hatches generally have controls allowing them to be operated from inside or outside of an airplane via quick-opening compression devices around the hatch's circumference. When released from the aircraft, hatches are likely to fall free from their own weight.

In a rescue situation, remember that people may be inside the aircraft next to the escape hatch. The hatch may also be blocked by cargo or freight located in the inside of a **cargo aircraft**, next to the escape hatch. If oxygen masks have been deployed, the hoses and small hinges mask drop-open doors that have dropped open potentially posing an obstruction at the top edge of the hatch. (This depends on the aircraft type and presents a greater problem in smaller passenger and corporate type aircraft.)

Once the hatch has been unlocked and removed from the sidewall of the fuselage, it should be stowed in an area that will not obstruct passenger evacuation or hinder operations from a wing. Hatches, especially plug-style hatches, are almost

Figure 2-9
Firefighters entering a cargo jetliner through the standard entry door. The ladder has been placed to the side of the door opposite the hinge, and the area immediately below the door is kept clear for safety.

Danger
If you indiscriminately toss or drop an escape hatch from the wing of a grounded aircraft, the hatch may strike people standing in the proximity of the wing.

impossible to open if the aircraft is still pressurized (although it is unlikely that this will be the case).

If you indiscriminately toss or drop an escape hatch from the wing of a grounded aircraft, the hatch may strike people standing in the proximity of the wing (e.g., rescue personnel or evacuating passengers). Hatches may weigh as much as 45 pounds.

Figure 2-11 shows emergency responders gaining entry through a plug-style, over-wing hatch located on a commercial aircraft. In this training exercise, the main entry door escape slide is armed and still attached to the bottom hatch sill. Because of this, the entry of choice is the over-wing emergency escape hatch.

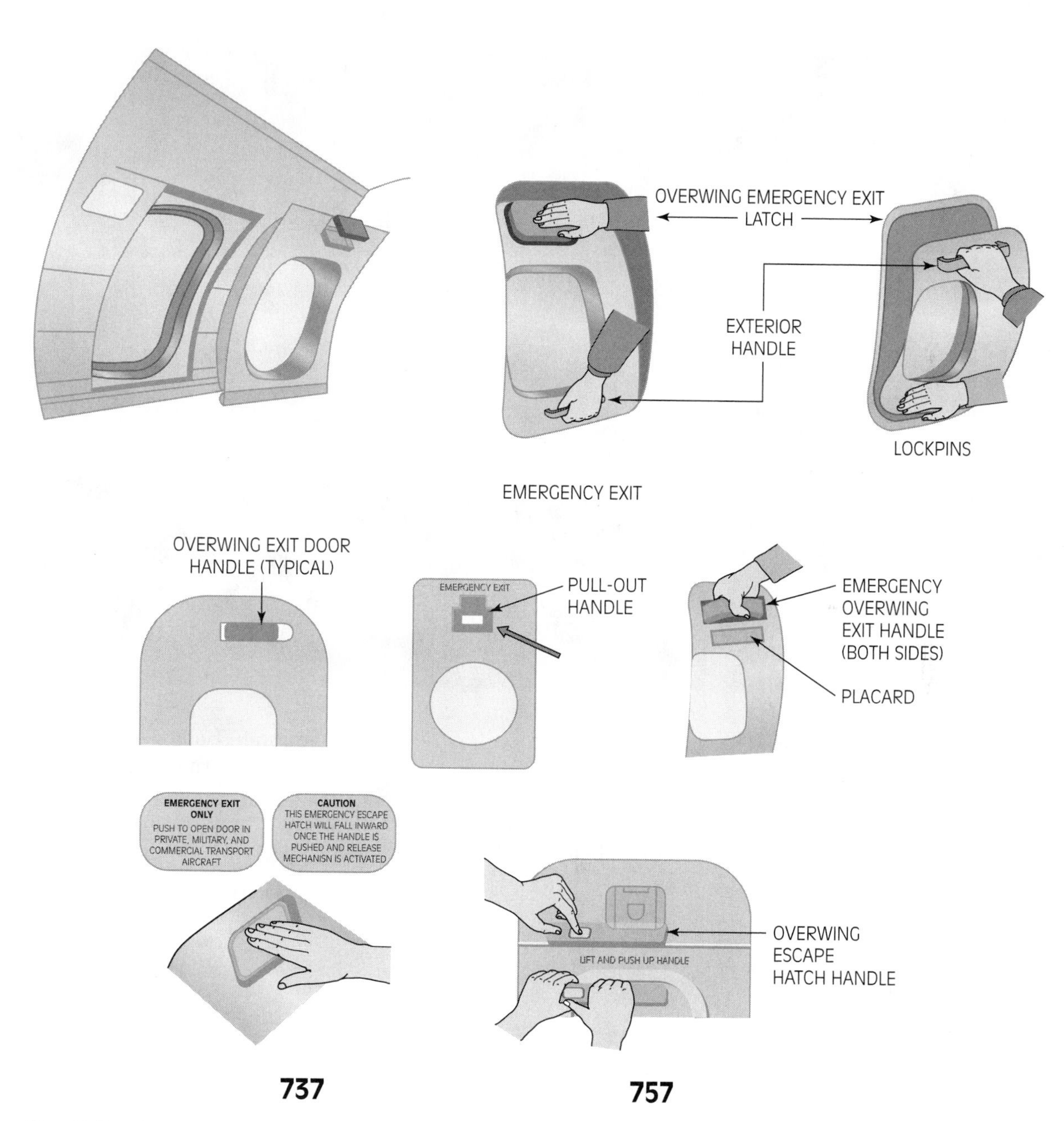

Figure 2-10 *An assortment of types of emergency escape hatches used in private, military, and commercial transport aircraft. (Courtesy of the U.S. Air Force.)*

Figure 2-11 *Rescuers gaining entry to a commercial aircraft through a plug-style, over-wing hatch.*

Hatches locations vary according to the type of aircraft. Hatches designed for *normal* personnel access are hinged and may be opened internally or externally. Most aircraft use plug-style emergency escape hatches, as shown in Figure 2-11.

The diagrams in **Figures 2-12** and **2-13** illustrate a turboprop airliner with its components clearly labeled.

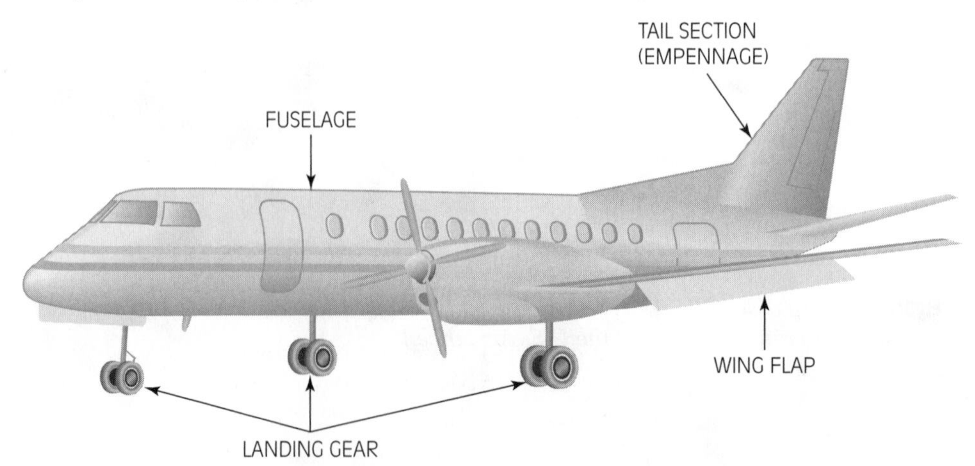

Figure 2-12 *A Saab 340 passenger plane, left view. (Courtesy of Dave Armour.)*

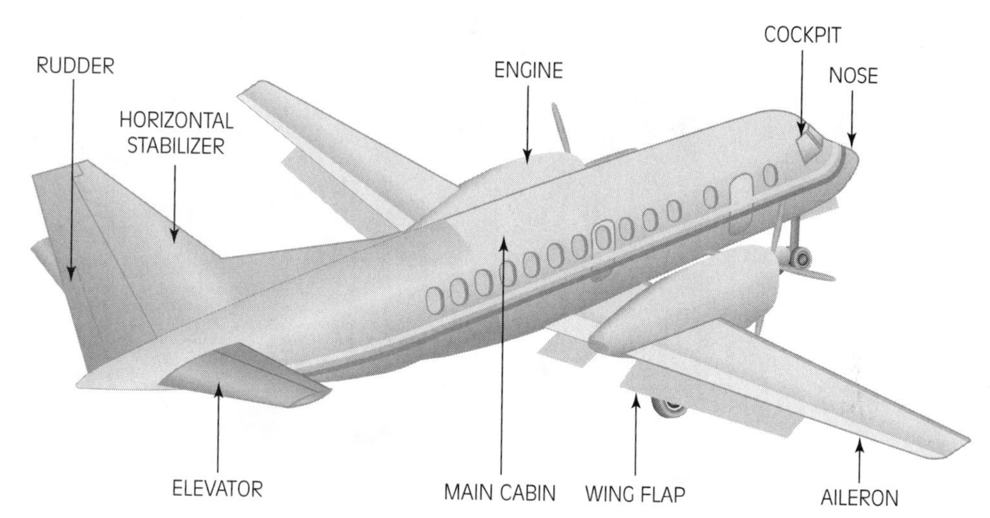

Figure 2-13 *A Saab 340 passenger plane, right view. (Courtesy of Dave Armour.)*

jetliner
jet-powered cargo or passenger-carrying aircraft

controls
any instruments or components provided to enable the pilot to control an aircraft's speed, direction of flight, altitude, power, and so on

firewall
bulkhead separating two compartments of an aircraft; for example, the engine compartment and the aircraft's cockpit/cabin

general aviation aircraft
all civil aviation aircraft used for private, unscheduled, non-revenue operations

The **empennage** is the complete tail assembly of an airplane, including its parts, or components, which are commonly the horizontal and vertical **stabilizers** of an aircraft. Reference the elevator and vertical stabilizer in Figures 2-12 and 2-13: An **elevator** is the movable control surface located on the horizontal stabilizer. This controls the upward and downward pitch of the aircraft. The vertical stabilizer is much like a fin. It houses the **rudder** that controls back-and-forth movements (called *yaw*). Some large-frame **transport aircraft** are equipped with a fuel tank located in the vertical stabilizer.

That portion of aircraft wing that is fastened to the fuselage is called the **wing root**, as shown in **Figure 2-14.** This is the attachment point where the wing bears the greatest amount of force (loading). Note the wing attachment points in Figure 2-14. The skeletal system, or framework, of a typical aircraft wing consists of **spars**, which are the framework running from wing tip to wing root, and **ribs**, which run from the front edge to the rear (**trailing edge**) of the wing.

Refer back to Figure 2-13, which also shows an **aileron**, a movable control surface located on each wing that enables the airplane to bank left or right. (This maneuver is similar to when you are riding a bicycle and you bank left or right to execute a turn.) Ailerons are hinged to the rear of the wings, usually outboard, toward the wing tips.

An airplane is more easily controlled at slower speeds due to **flaps**. Flaps are control surfaces hinged to the rear edges of the wings that provide an aircraft with additional lift at slower speeds. When flaps are left in the extended out and down position, they enhance passengers' escape during an emergency evacuation. Attached to the **leading edge** of the wings are extendable **slats**, which, like flaps, provide better lift and control of the airplane at slower speeds. Not all aircraft are equipped with leading edge slats.

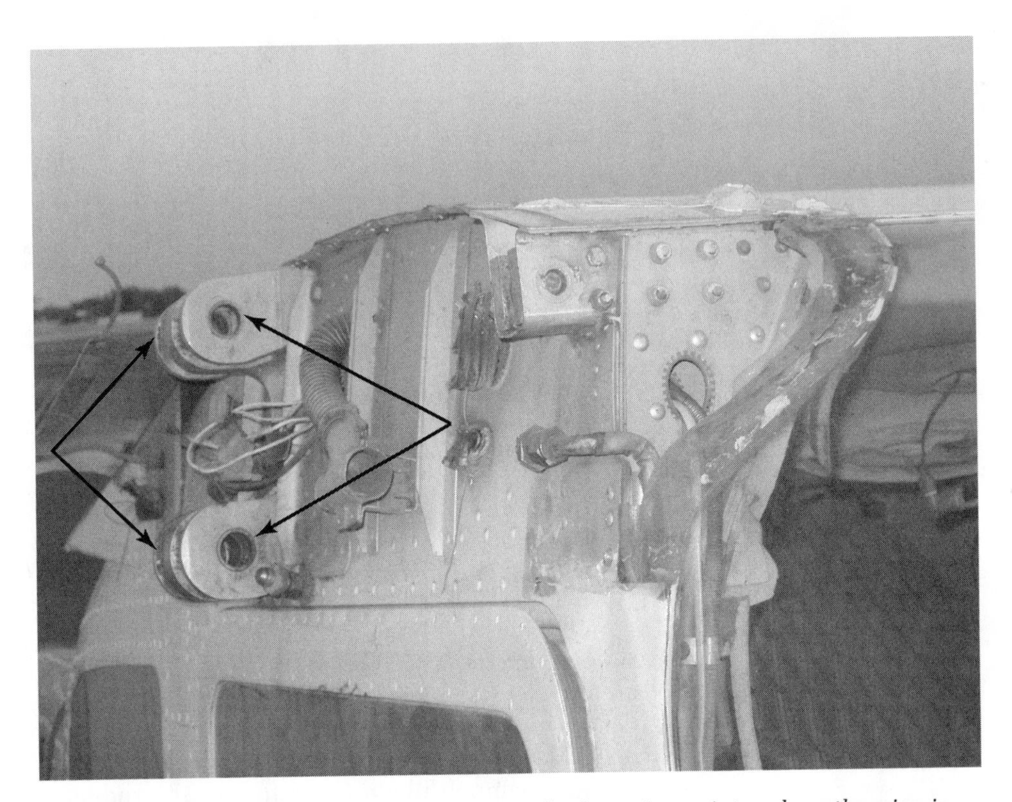

Figure 2-14 *The wing root: The arrows indicate the fastening points, where the wing is connected to the fuselage.*

airliners
in the United States, an aircraft designated primarily for the transport of paying passengers

emergency evacuation
rapid exiting of an airplane during a situation posing threat of bodily harm or death

military aircraft
any kind of airplane owned and operated by military forces

A wing strut, shown in **Figure 2-15,** is connected between the bottom of the wing and the bottom-side area of the fuselage to strengthen the aircraft.

The passenger jetliner pictured in **Figure 2-16** shows jet engines attached to the aft (rear portion) of the fuselage.

Unless attached to the front or rear of an aircraft, engines are mounted to the wings with pylons. The aircraft in the bottom of **Figure 2-17** has pylon-mounted engines. The diagrams show a Boeing 717 (in the DC-9/MD-80 family; top) and a Boeing 757 (bottom). Looking at Figure 2-17, you can also see the auxiliary power unit (APU), engine, and fuel line locations.

Auxiliary power units (APU) are fuel-powered (usually turbine) units that supply electrical power, air conditioning, and backup power to an aircraft during flight. The APU may also be used to power pneumatic (air) and hydraulic (fluid) pumps within the airplane. Some fighter aircraft, such as the F-16, use an emergency power unit (EPU) instead of an APU. In the event of engine failure, the EPU, which is powered by a toxic fuel called **hydrazine**, automatically starts and furnishes power for flight instruments and aircraft control movements.

Figure 2-15 *A wing strut attaching the wing to the fuselage of a Cessna 172.*

Figure 2-16 *The jet engines of this Embraer 145 regional jetliner are attached to the aft (rear portion) of its fuselage. The empennage is in what is commonly called a "T tail" design.*

hatches
openings that provide a means for escape from, and emergency entry into, a distressed aircraft

Figure 2-18 shows a digital **flight data recorder (FDR)**, mounted in the rear fuselage of an eight-passenger Gulfstream business (corporate) jet aircraft. This machine records in-flight information such as airspeed, aircraft flight attitude, and engine RPM, and can record outside air temperature; vertical acceleration, other flight variables. This is critically useful data for accident investigators. FDRs may also be referred to as a *black box*, *flight recorder*, or *digital flight data recorder*.

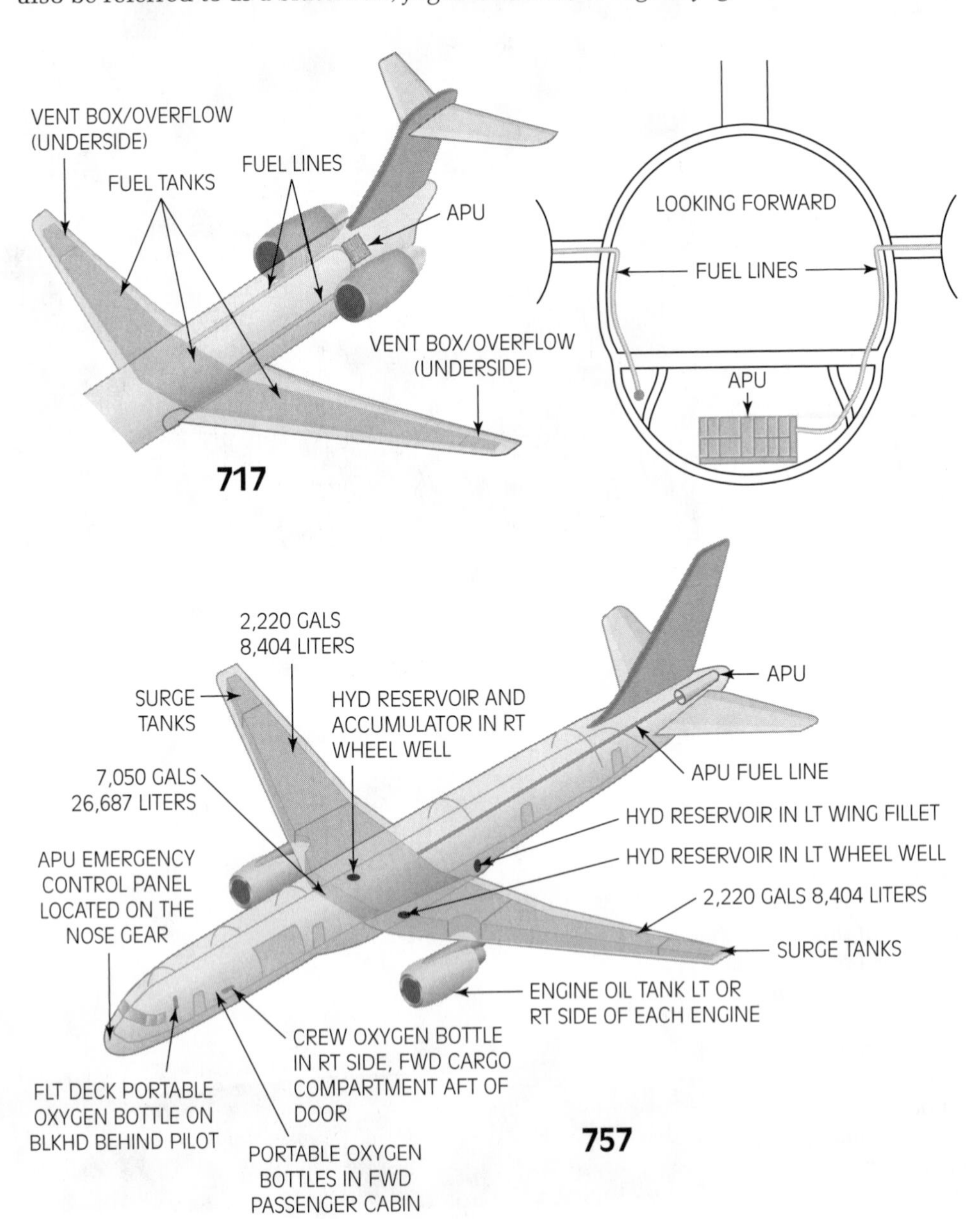

Figure 2-17 *A diagram illustrating APU locations on a Boeing 717 and 757. Note the locations of fuel lines; if you have to cut into an aircraft, it is important to know where the fuel lines are.*

figure 2-18 *A flight data recorder located in the tail (aft) area of the fuselage.*

commercial aircraft civil aircraft; most commonly used for transportation of people and freight for revenue on a scheduled or charter basis

cargo aircraft designated to carry freight; may be civil or military aircraft

empennage complete tail assembly of an aircraft and its parts or components, including the horizontal stabilizer, elevators, rudder, and so on

AIRCRAFT SIZE AND WEIGHT

Aircraft are categorized by payload size. In commercial passenger transport aircraft, classification also is according to fuselage width and the number of passenger seats.

Aircraft Weight Classes

Small aircraft are defined as having 41,000 pounds or less, maximum, of certificated takeoff weight. Examples of small aircraft include the Beech 1900, Bonanza, Super King Air 350, Cessna 172, Cessna Citation Jet, and Piper Apache 150/160. Some aircraft in this class carry small numbers of people and/or amounts of cargo payload.

Heavy aircraft, sometimes nicknamed *jumbo jets*, are capable of accommodating takeoff weights of more than 255,000 pounds. These aircraft are designed for long-distance flights. Examples of heavy aircraft include the Airbus A300, 330, 340, 350, and 380; Boeing 747, 767, 777, and 787; B1; DC-10/MD-11; B-1, B-2, and B-52 bombers; KC-10, C-5, C-130, and C-141.

Figure 2-19 *A Boeing 757. (Courtesy of Dave Armour.)*

stabilizers
components that house the rudder that controls back-and-forth movements (called "yaw")

elevator
movable horizontal portion of the tail of an aircraft; hinged to the rear of the horizontal stabilizer and controlled by the pilot to move the aircraft's nose up or down or to level its flight position

Large aircraft are capable of hauling more than 41,000 pounds, maximum, of certificated takeoff weight. Weight is up to 255,000 pounds. Examples of large aircraft include the Airbus A319 and 320; Boeing 717, 727, 737, 757, DC-9, MD-80, Bombardier CRJ, and Embraer 145 Regional Jets; F-15, F-16, F-18, F-22, and F-35. **Figure 2-19** shows the Boeing 757, a popular aircraft for passenger as well as cargo operations.

(Note: The FAA takes an additional factor into consideration when classifying [or "indexing"] aircraft for airport use: the length of an aircraft's passenger-carrying fuselage. This factor is discussed more extensively in ARFF training courses.)

Most Boeing 727s have been replaced by a new generation of jets for passenger and cargo flights. The sturdy, venerable 727 has proven itself as a versatile

Figure 2-20 *The lower cargo hold of a passenger jetliner.*

rudder
upright, movable part of the tail assembly, which controls the direction of the aircraft

transport aircraft
aircraft designated for the purpose of carrying passengers or cargo (freight)

wing root
point at which an aircraft wing is joined to the fuselage

spars
principal structural members, or beams, of a wing

and reliable airplane for long-range passenger and cargo operations, and a few are still in service, but its numbers are steadily decreasing. The new passenger jetliners, such as the Embraer 170 and 190, Airbus A350 and A380, and Boeing 787, are capable of transporting large payloads of cargo or people for greater distances more efficiently and with less fuel consumption.

Most passenger-carrying commercial aircraft transport freight in addition to baggage. As shown in **Figure 2-20,** below the main deck is another deck, called the cargo hold or baggage hold. This confined space is not intended for human occupation and has only one point of entry/exit.

TYPES OF AIRCRAFT

General Aviation Aircraft

General aviation (GA) aircraft include a variety of sizes and styles of airplanes that normally are used for private sport, leisure, or business flying. **Figure 2-21** shows a Beech Bonanza, a popular private aircraft, parked on the ramp at a

Figure 2-21 *A Beech Bonanza private aircraft. Note that the empennage, which is referred to as a "V tail," is different from those of most aircraft.*

ribs
part of the skeletal structure of an aircraft wing that gives the airplane form, strength, and shape

trailing edge
rear edge of an airfoil; applies to tail surfaces, wings, propeller blades, and so on

aileron
a control surface that consists of a moveable, hinged portion of an aircraft wing

municipal airport. This small-frame airplane, like others in its category, possesses the same basic structural components as larger commercial or military airplanes, although the design can differ as widely as designs for automobiles. Most of these private aircraft are small-frame (such as the Cessna 172; Beech; Piper; Cirrus; Lear 21, A-37, and T-37; and Mooney 201) and contain piston, turboprop, or jet-powered engines.

Some larger private airplanes, such as the Hawker Horizon jet, are considered medium-frame aircraft. Some large-frame commercial jets belong to private corporations and have undergone major interior design modifications, such as the installation of customized seating, dining, conference, and office furnishings. Other modified corporate jetliners are outfitted with additional luxury amenities such as sleeping accommodations.

Figure 2-22 shows the cockpit of a privately owned, twin-engine business jet in the general aviation aircraft family. **Figure 2-23** provides a view of the passenger cabin in this same airplane. You can easily see that a rescue from this plane would be difficult because of the small area inside the fuselage.

As you learn more about aircraft and the special firefighting and rescue needs associated with them, you will discover their common construction features: Fuel is carried in the wings, and all have propellers, airframes, fuselages, and

Figure 2-22 *The cockpit of a six-passenger private jet.*

Figure 2-23 *The passenger seating area of a six-passenger private jet.*

flaps
adjustable airfoils that are attached to the leading or trailing edge of the wings, affecting the aircraft's aerodynamic performance during landings and takeoffs

leading/trailing edge
forward/rear edge of an airfoil; applies to tail surfaces, wings, propeller blades, and so on

electrical systems. The illustrations in this book will help you identify additional characteristics that aircraft share, as well as crucial differences among them.

Small-Frame Aircraft The small-frame category of GA aircraft includes single-engine planes, some twin-engine aircraft, very light jets, and commuter airplanes.

VLJ Aircraft **Very light jets** (**VLJs**, sometimes called **microjets**) are aircraft that use low-noise turbofan jet engines. VLJs are inexpensive compared to other jet aircraft, and many have a cruising speed of 325–375 mph, with maximum flight altitudes of nearly 40,000 feet. These aircraft are capable of taking off from runways as short as 3,000 feet, which are common at small private and municipal airports.

VLJs' ability to land on short runways has increased public access to *air-taxi service*, especially to underserved communities, for business and leisure travel. Historically, this service was used almost exclusively for private business travel, but the industry has been growing rapidly since September 11, 2001, due to changes in the commercial air travel industry (such as security, aircraft size, and the need to move more people between airports). Most air-taxi operators fly small aircraft ranging in size from the popular single-engine Cessna 172 to the

slats
movable auxiliary airfoils whose primary function is to increase the aircraft's stability; usually references the fixed horizontal tail surface of the aircraft

hydrazine
fuel used in the emergency power unit on aircraft such as the F-16; extremely toxic and caustic

twin-engine, piston-powered Beech Baron and Piper Aerostar to business jets such as the Lear, Raytheon/Beech, Hawker, and Gulfstream aircraft.

Commuter Aircraft Depending on their size and design, **commuter aircraft** can carry up to nineteen passengers. Often nicknamed *puddle jumpers*, these usually single-engine aircraft meet an important traveling need: Like the regional airliners, they fly from small cities to medium and large cities' airports, thus connecting small cities with the major commercial airlines' "hubs" in big cities. Common examples of commuter aircraft are the Beechcraft 1900, Jetstream 31 and 41, and Embraer 120. Older commuter planes, such as the Fairchild Metroliner, now are used to carry cargo from small airports to medium or large airports. Some twin-engine aircraft, such as the Cessna 421, are used as commuter aircraft.

All of the aircraft discussed above (except military bombers and **combat aircraft**), are also used for air cargo conversions. A few airlines are still using the large four-engine L-188 Electra for cargo operations.

Figure 2-24 displays a small twin-engine aircraft that has been modified to be a cargo plane. This aircraft carries freight on short-distance routes, or *feeder routes*, transporting cargo shipments to larger airports where they are transferred

Figure 2-24 *A general aviation, twin-engine aircraft converted for cargo use.*

to larger aircraft. As we study the exterior view in this picture, we notice that access to the pilots is through the left front side window, dangerously close to the propeller.

Figure 2-25 shows the inside of this converted aircraft. Note that the main cargo compartment of the fuselage is isolated from the flight deck by a solid bulkhead.

flight data recorder (FDR)
device that records in-flight information such as speed, engine RPM, aircraft flight attitude, pitch, and so on; also can record outside air temperature, vertical acceleration, and other variables while an aircraft is in the air

small aircraft
41,000 pounds or less takeoff weight; carry small amounts of people and/or payload

heavy aircraft
can accommodate takeoff weights of more than 255,000 pounds; designed for long-distance flights

Medium-Frame Aircraft Larger than commuter airliners, and with a seating capacity of less than 100 passengers, medium-frame **civil aircraft** are classed as *regional airliners*. Normally designated for shorter flight routes, regional airliners serve communities that do not have the demand for wide-body aircraft. Often, these "regionals" connect passengers from smaller cities with airline hubs at large airports. Common examples of this class of aircraft include the Bombardier CRJ series and Dash 8, Embraer Regional Jets, Saab 340 and Saab 1000, and ATR 42 and 72 turboprop.

Figure 2-26 displays the Brazilian-built Embraer 145 series of regional jets, which, with the popular Canadian-built Bombardier CRJ aircraft, are used extensively for commuter and longer-range regional routes. The Embraer 135/145 series carries between thirty-seven and seventy passengers, depending on the airplane's configuration. These aircraft also are useful on routes that previously were too costly for older or larger aircraft to operate.

Figure 2-25 *An inside view of the cargo compartment of the airplane in Figure 2-24.*

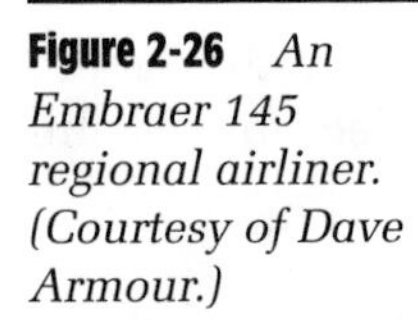

Figure 2-26 *An Embraer 145 regional airliner. (Courtesy of Dave Armour.)*

large aircraft
of more than 41,000 pounds, maximum certificated takeoff weight, up to 255,000 pounds

very light jets (VLJ)
aircraft that use low-noise turbofan jet engines; sometimes called "microjets"

The Bombardier Dash 8 family of aircraft shown in **Figure 2-27** is used worldwide for commuter and regional passenger service. This aircraft has a reputation for great performance and reliability, and has **short take off and landing (STOL)** abilities, making it ideal for use at smaller airports.

Large-Frame Aircraft Large-frame aircraft include narrow- and wide-body airplanes.

Narrow-Body Aircraft Narrow-body aircraft are larger than regional airliners. These aircraft generally are used for medium-distance flights and have one aisle.

Wide-Body Aircraft The largest airliners are wide-body jets, which fall into the heavy aircraft classification. Passenger aircraft in this category are commonly

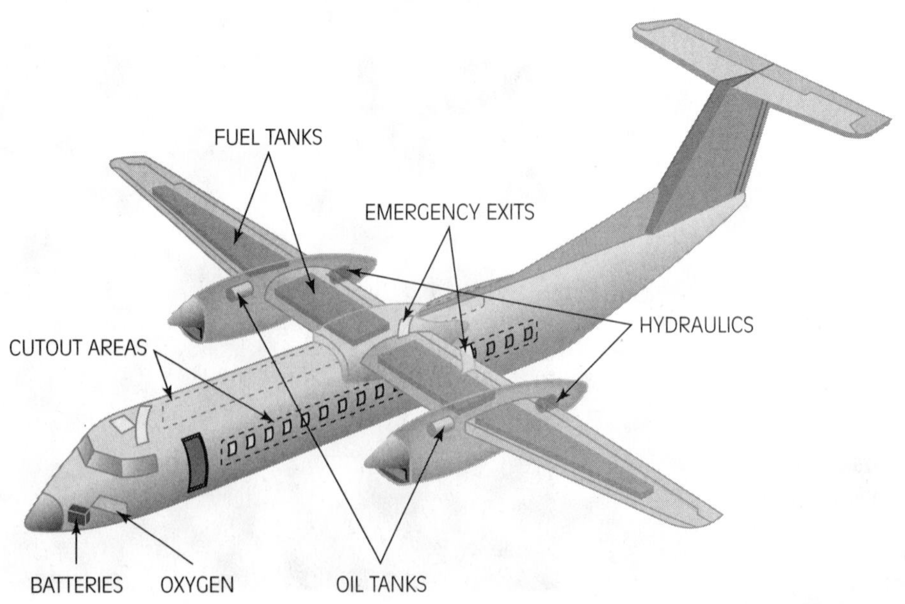

Figure 2-27 *A Dash 8 regional passenger airliner. (Courtesy of Lahey Applications and Design Services.)*

microjets
a class of aircraft called very light jets (VLJs)

commuter aircraft
small- or medium-frame aircraft that flies passengers on short routes

combat aircraft
military aircraft that are designated for use in warfare and can be carrying weapons systems

civil aircraft
two categories of non-military aviation: general aviation (private) and commercial

referred to as twin-aisle aircraft because most of them contain two aisles in the passenger (main) cabin.

Military Aircraft

Like commercial and general aviation aircraft, military aircraft configurations are diverse. Military aircraft are used for purposes such as combat, weapon delivery, troop transfer, disaster relief, and cargo transport. The design of these aircraft is primarily a function of purpose: an airplane's intended mission determines its design. For example, the F-22 is designed for air-to-air combat, while the C-17 is configured for transport/cargo duties in combat operations. The military cargo transport aircraft of most nations feature a mixed cargo-passenger configuration. Other military aircraft, such as the Boeing 737 family, are commercial aircraft purchased from the factory and modified to meet military mission needs. The KC-10, for instance, was purchased and modified for aerial refueling assignments.

A large military aircraft with impressive abilities is the C-17 Globemaster III, shown in **Figure 2-28.** Built specifically to meet the demanding rigors of military

Figure 2-28 *A C-17 Globemaster III military transport aircraft.*

short take off and landing (STOL)
designed to take off or land on short runways

bomb bay
an enclosure in an aircraft fuselage with doors that open when bombs or other weapons are being released

ordnance
explosive weapons, including ammunition, bombs, and rockets, that may be carried on military aircraft

operations, and able to land on and take off from short runways, this aircraft has proven itself as a great cargo transport in day-to-day military shipping and on disaster relief missions. Whether its mission is carrying cargo or relief supplies, or evacuating people, this aircraft gets the job done.

Many combat aircraft have a compartment located inside the lower aircraft fuselage called a **bomb bay**. A bomb bay (or *weapons bay*) is equipped with doors on the underneath side of the fuselage that are specially designed to open to release bombs or other weapons while the aircraft is in flight. **Figures 2-29** and **2-30** show bomb-bay doors on firefighting aircraft. These doors are virtually identical to those on combat aircraft such as attack planes, bombers, and fighters.

The ammunition, bombs, explosives, or rockets carried on combat aircraft are often referred to as **ordnance**. (This subject is discussed in more detail in Chapter 3, as well as in the tactical considerations sections in this book.) Emergency responders to military aircraft crash sites face complex situations, because they must be prepared to encounter any kind of cargo and/or baggage—including weapons and explosives—as well as a varying number of passengers on board.

Figure 2-29 *A DC-4 firefighting bomber.*

Figure 2-30 *A P-3 fire-retardant bomber equipped with bomb-bay doors.*

Military aircraft, like civilian aircraft, can be small-frame; these usually are fighter or training aircraft, such as the T-6, T-38, F-16, F-18, and A-10. Large-frame military aircraft, such as the C-17, C-5, and KC-10, are used for transport and often are called the "haulers" of the military services.

Specialized Aircraft

specialized aircraft used for duties not found in mainstream commercial or military aviation

Many emergency responders are unaware of the assortment of **specialized aircraft** used for diverse duties such as farming (agricultural spraying and seed planting), aerial firefighting, weather data gathering, drug interdiction, law enforcement, medical evacuation, and other tasks not found in mainstream commercial or military aviation.

agricultural aircraft dispense products such as fertilizers, pesticides, herbicides, and seeds

Farming from the Sky **Agricultural aircraft** account for 25 percent of crop protection activities in the United States. These aircraft dispense farm-related products, such as fertilizers; pesticides; and herbicides in liquid, powdered, or granular form, as well as plant seeds. Modern-day farmers feed many more people than were fed decades ago. Fewer people in the world are going hungry, thanks to developments in agriculture and this would impossible without civil aviation.

Figure 2-31 *A Thrush model agricultural aircraft.*

Like military and airline pilots, agricultural pilots are well-trained professionals; they routinely fly their planes safely and successfully into areas that are often hard to access, circumnavigating obstacles such as trees, buildings, and power lines with precision.

Some specialized agricultural duties, such as spraying crops to protect them from damaging insect infestations, are carried out by converted general aviation aircraft. **Figure 2-31** shows a Thrush airplane, which is built for agricultural purposes. Many helicopters also are used for agricultural purposes.

Flying Fire Trucks Another specialized aircraft is the aerial firefighting aircraft commonly referred to as a *tanker*. Pilots of these aircraft, like their counterparts in military and agricultural aviation, skillfully fly their aircraft in the adverse conditions, such as close proximity to harsh terrain and through smoke, extreme heat, and the unstable air currents above large fires. Some aircraft tankers refuel other aircraft in flight, but the aircraft referred to here are flying fire trucks: They deliver and expel fire-retardants. Refer back to Figure 2-29, which shows a DC-4 airliner modified for firefighting duty after years of service as a passenger-carrying and cargo aircraft. Note the bomb-bay doors on the bottom of the fuselage: These doors are identical to those on a combat aircraft and enable the delivery of firefighting agent. When pilots are ready to release an agent (water or a chemical slurry, described in the next paragraph), they open these doors and release the agent.

Fire-retardant and -suppression materials have changed over the years. Those dropped in the past contained materials that inhibited the regrowth of plants and polluted water, resulting in the death of marine life. Today, firefighting aircraft drop ammonium sulfate or ammonium polyphosphate mixed with a material that thickens the retardant, such as clay or guar gum. Commonly used commercial brands of aerial fire retardants include Fire-Trol and Phos-Chek, and these names are often used in place of the term *fire retardant.* Some aircraft drop an application of water mixed with a wetting (penetrating) agent that promotes penetration of thick vegetation. Just as with ammonium sulfate or ammonium polyphosphate, a thickening agent may be added to the water to enable it to cling to vegetation.

As an emergency responder, you need to know these facts, because today's firefighting agents, although less toxic than earlier versions, still pose dangers. For instance, these materials can be slippery, especially on sloped surfaces. If an aircraft carrying these materials crashes on a slope, any spilled retardant may make the ground unnavigably slick, slowing your efforts to reach the stricken aircraft.

Several types of aircraft are used as firefighting tankers. Figure 2-30 shows a P-3 fire-retardant bomber. For decades, military patrol aircraft version of this airplane, the P-3 Orion, successfully served many nations. The commercial version of this aircraft, the Lockheed L-188 Electra, may be used for carrying passengers and freight. Like other firefighting tankers, this aircraft also has bomb-bay doors on the bottom of the fuselage. Other aircraft modified for firefighting duties include the C-130, S-2 Tracker, and PB4Y Privateer.

■ Note
Other aircraft modified for firefighting duties include the C-130, S-2 Tracker, and PB4Y Privateer.

Jet aircraft are entering the firefighting scene with the Russian-built Ilusion IL76 four-engine cargo conversion, as well as the Boeing 747, DC-10 Tri-Jet conversions, which were used during the 2006 wildland fire season in California. These "super tanker" aircraft have proven to be effective over most terrain, but are not yet in wide use. As of this writing, their cost effectiveness is being closely studied. Another aircraft in this category is the Canadair Bombardier. CL-215, which is amphibious, meaning that it can take off from or land on water. It can be refilled with water (1,620 gallons' worth) by skimming the surface of large bodies of water, such as lakes. Although it was designed and built solely for the purpose of firefighting, it is used in some countries for search-and-rescue duties.

The helicopter in **Figure 2-32** is the strong and versatile Sky Crane, which was a workhorse during the Vietnam War. Known for its versatility and ability to carry huge, heavy payloads. Currently, one of the missions of these friendly giants is use as firefighting helicopters. They are able to resupply themselves with water by dropping a large hose into an on-ground water container and pumping the water into its tanks.

Figure 2-32 *A Sky Crane firefighting helicopter.*

THE SIMILARITIES AND DIFFERENCES AMONG AIRCRAFT

fixed-wing aircraft
consist of a fuselage, wings, and a tail assembly

ultralight aircraft
characterized by a small airframe and motor; designed for recreational purposes and do not require an FAA registration number, airworthiness certificate, or pilot certification

All **fixed-wing aircraft** have the same basic components: a fuselage, wings, a tail assembly (empennage), wheels, and wheel (landing gear) struts. They also carry fuel and have one or more power plants (engines), as well as some form of electrical system. In addition, most aircraft, unless they are very small (such as an ultralight, a Piper J-3 cub, or a Cessna-type airplane with fixed landing gear), have some type of hydraulic system. Important differences exist among aircrafts, however.

Ballistic Recovery Systems

Some aircraft, however, such as the Cirrus, some Cessna, and some **ultralight aircraft**, may be equipped with a **ballistic recovery system (BRS)**. This life-saving device, which acts as a parachute for an airplane in trouble, also, ironically, is a life-endangering one to anyone who must attempt to disable it in a crashed airplane. The BRS contains explosive charges and has hatch covers that explosively blow away (like the jettison/egress system in some military aircraft). When a ballistic recovery system is activated, a rocket drags a tightly compacted parachute up and toward the rear of the aircraft. Currently, warning decals for this device are extremely small. The activating cable within the system is manually deployed; the pilot pulls the activating handle, which requires approximately 35 pounds of force.

Danger When a ballistic recovery system is activated, a rocket drags a tightly compacted parachute up and toward the rear of the aircraft. Currently, warning decals for this device are extremely small.

A BRS-equipped aircraft whose airframe has been severely bent may be extremely dangerous, because the bent airframe can result in a potentially unstable BRS system. (An airframe can be bent during ground collisions, during a ground collision while taxiing, or as the result of forces of nature, such as a tornado.) The bent airframe may have stretched the system's activating cable, placing it under tension. Use extreme caution during aircraft stabilization and occupant rescue/extrication procedures on BRS-equipped aircraft.

Danger A BRS-equipped aircraft whose airframe has been severely bent may be extremely dangerous, because the bent airframe can result in a potentially unstable BRS system.

Cutting the activation cable for the BRS *must* be done with a Felco or other battery-cable cutter (Felco is a brand name; Felco and other battery-cable cutters are available from most automobile parts stores). Cut the activating housing at the base of the launch tube, where the housing screws onto the tube. When cutting, be careful not to twist the cable housing, as this may initiate the system. (Call Cirrus Aviation if you are in doubt about the effectiveness of any battery-cable cutter for use on a BRS.) After the cable has been cut, you may elect to remove the still-live rocket motor to a secure place and transfer it to the appropriate aviation safety authorities when they arrive on scene.

The reason that BRS cable *must* be cut with a Felco or other battery-cable cutter is that if bolt, wire, or ejection-seat-catapult hose cutters are used, they will pull on the initiation cable with enough force to fire the mechanism. The system's parachute is tightly packed into its container, weighs approximately 55 pounds, and has the density of a fireplace log when initially deployed. It accelerates at more than 100 miles per hour with more than 300 pounds of force. Because the parachute is pulled up away from the airplane, it can severely injure and may kill anyone in its path. In addition, anyone standing close to the airplane may be struck with shrapnel caused by the support straps pulling away from the fuselage sides during parachute deployment.

Danger The reason that BRS cable must be cut with a Felco or other battery-cable cutter is that if bolt, wire, or ejection-seat-catapult hose cutters are used, they will pull on the initiation cable with enough force to fire the mechanism.

If the airplane's frame is badly distorted or resting at an unnatural angle, it is *not possible* to predict in which direction the BRS parachute will deploy.

For safety reasons, you should always thoroughly examine an accident scene, but particularly when assessing a crashed Cirrus or Cessna aircraft, be sure to determine whether it is equipped with a BRS. If so, the aircraft should display a warning decal on the rear cabin windows. In addition, look for a transparent, hollow plastic tube at the rear of the passenger cabin, with what appears to be a pressure vessel at its base—if you see such a device, the airplane has a BRS. Over time, this technology is likely to be embraced by additional aircraft manufacturers.

ballistic recovery system (BRS)
emergency device that consists of explosive charges and hatch covers that, once activated, fire a rocket that drags a tightly compacted parachute up and toward the rear of the aircraft, enabling the airplane to float safely to the ground

Other Differences

Some specialized aircraft, such as those used by the National Oceanic and Atmospheric Administration (NOAA), the Federal Aviation Administration, and law enforcement agencies, may be outfitted with specialized imaging systems. These systems include, but are not limited to, *forward-looking infrared radar* (FLIR).

Law enforcement airplanes may carry small weapons and ammunition. Some general aviation airplanes have been modified to spray pesticides on crops or control disease-carrying insects, such as mosquitoes. These chemicals can be toxic if inhaled or if they come in contact with skin. Specialized research and testing aircraft used by the National Aeronautics and Space Administration for testing and research may be general aviation, commercial, or converted military aircraft.

Military combat aircraft, given the nature of their mission, contain weapons delivery systems, munitions, exotic radar imaging systems, and egress/jettison systems, which is discussed in detail later in this book.

Like the DC-10/KC-10, some aircraft in use by the military are civilian aircraft that have been purchased for military use and remain "as is" in their design layout. These include the C-12, C-21, C-32, C-37, and C-40. Other civilian aircraft, however, have been modified to suit military needs, such as the E-3 and KC-135 (Boeing 707), KC-10 (McDonnell Douglas/Boeing DC-10), E-4, VC-25, and YAL 1A (Boeing 747).

SUMMARY

By now, you should have a basic understanding of the anatomy and purpose of various types of military, general aviation, and commercial aircraft.

- All fixed-wing aircraft have a fuselage, wings, fuel, and means of propulsion (engines).
- All aircraft have electrical systems, airframes, and batteries.
- Most aircraft (except the very small) have hydraulic systems, which pose additional threats to responders.
- Dealing with a crashed combat aircraft, such as an A-10 or F-22, poses a wide range of hazards different from those of a crashed commuter or cargo aircraft. Combat aircraft may carry explosive munitions. (You will learn later in this book how to safely approach these aircraft.)
- Similar aircraft types may be designated to carry out vastly different missions, such as weather data gathering, firefighting, cargo, or passenger transport. An example is the P-3 Patrol aircraft, which is used by the military, and the P-3 aerial firefighting tanker aircraft.
- Preplanning includes understanding not only what kinds of aircraft traverse your community's airspace, but also these airplanes' basic anatomy and the functions of their various systems.
- Check your progress by completing the following exercises. Once they are completed, you will be ready to discover more about aircraft hazards in Chapter 3.

KEY TERMS

Agricultural aircraft These dispense products such as fertilizers, pesticides, herbicides, and seeds.

Aileron A control surface that consists of a moveable, hinged portion of an aircraft wing. Ailerons usually are part of the trailing edge of a wing. Their primary function is to help tilt the wings when an aircraft is banking.

Aircraft weight classes:

- **Heavy aircraft** can accommodate takeoff weights of more than 255,000 pounds; designed for long-distance flights.
- **Large aircraft** of more than 41,000 pounds, maximum certificated takeoff weight, up to 255,000 pounds.
- **Small aircraft** 41,000 pounds or less takeoff weight; carry small amounts of people and/or payload.

Airframe Generally, the parts of the aircraft having to do with the flight: the fuselage, boom, nacelles, cowlings, fairings, empennage, airfoil surfaces, landing gear, and so on.

Airliner In the United States, an airliner is defined as an aircraft designated primarily for the transport of paying passengers. These aircraft are usually operated by an airline that operates under FAA regulations for scheduled or charter airline transportation.

Auxiliary power unit (APU) This fuel-powered (usually turbine) unit supplies electrical power, air conditioning, and backup power to an aircraft during flight. The APU may also be used to power pneumatic (air) and hydraulic (fluid) pumps within the airplane.

Ballistic recovery system (BRS) This emergency device consists of explosive charges and hatch covers that, once activated, fire a rocket that drags a tightly compacted parachute up and toward the rear of the aircraft, enabling the airplane to float safely to the ground.

Bogie A tandem arrangement of landing gear wheels. Bogies swivel up and down to enable all wheels to follow the ground as the altitude of the aircraft changes or as ground surface changes.

Bomb bay An enclosure in an aircraft fuselage whose doors open when bombs or other weapons are being released.

Bulkhead An upright partition that separates one aircraft compartment from another. Bulkheads may carry a part of the structural stress while forming the shape of an aircraft fuselage; equipment and accessories may be mounted on them.

Cabin Passenger compartment in an aircraft.

Canopy The transparent enclosure over the cockpit on some aircraft.

Cargo aircraft These aircraft, also called *transport aircraft,* are designated to carry freight and may be civil or military aircraft. Some cargo style aircraft have been converted for private, or VIP, operations.

Civil aircraft This term refers to two categories of nonmilitary aviation: general aviation (private) and commercial. General aviation (GA) includes all civil flights, whether private or commercial.

Cockpit Compartment where the pilots sit to fly the aircraft.

Combat aircraft These military aircraft are designated for use in warfare and can carrying weapons systems. Examples include attack, bomber, electronic-warfare, and fighter aircraft.

Commercial aircraft The official name for this category of aircraft is civil aircraft. Civil aircraft are most commonly used for transportation of people and freight for revenue on a scheduled or charter basis.

Commuter aircraft This term describes a small- or medium-frame aircraft that flies passengers on short routes.

Control surfaces Ailerons, flaps, elevator, rudder, and spoilers, which control an aircraft's direction of flight, altitude, and pitch.

Controls Any instruments or components provided to enable the pilot to control an aircraft's speed, direction of flight, altitude, power, and so on.

Elevator The movable horizontal portion of the tail of an aircraft. The elevator is hinged to the rear of the horizontal stabilizer and controlled by the pilot to move the aircraft's nose up or down or to level its flight position.

Emergency evacuation The rapid exiting of an airplane during a situation posing threat of bodily harm or death.

Emergency power unit (EPU) Some fighter aircraft, such as the F-16, use an EPU instead of an auxiliary power unit. The EPU is powered by a toxic fuel called hydrazine, and, in the event of engine failure, it automatically starts and furnishes power for flight instruments and aircraft control movements.

Empennage An aeronautical term referring to the complete tail assembly of an aircraft and its parts or components, including the horizontal stabilizer, elevators, rudder, and so on.

Engine The motive that powers the airplane and allows it to travel. May be piston-driven propeller, turboprop, or jet engines.

Firewall A bulkhead separating two compartments of an aircraft, for example, the engine compartment and the aircraft's cockpit/cabin.

Fixed-wing aircraft Airplanes that consist of a fuselage, wings, and a tail assembly.

Flaps These adjustable airfoils are attached to the leading or trailing edge of the wings, affecting the aircraft's aerodynamic performance during landings and takeoffs. Usually extended during landings and takeoffs.

Flight data recorder (FDR) This device records in-flight information such as speed, engine RPM, aircraft flight attitude, pitch, and so on. It also can record outside air temperature, vertical acceleration, and other variables while an aircraft is in the air. It may also be referred to as a *flight recorder* or *digital flight data recorder*.

Flight deck The pilot's compartment of a large airplane. Also referred to as a *cockpit*.

Fuselage This is the main body of the aircraft, to which the wings and tail are attached. Usually, construction incorporates frames and bulkheads that make up the individual compartments or sections of an airplane and strengthen the fuselage.

General aviation aircraft All civil aviation aircraft used for private, unscheduled, non-revenue operations. The most common examples are Cessna, Piper, and Beech. This category also includes lighter-than-air balloons, air ships, sail planes, gliders, rotorcraft (helicopters), and gyroplanes. The majority of all aircraft flights in the United States involve general aviation aircraft.

Hatches Openings that provide a means for escape from, and emergency entry into, a distressed aircraft. They may be located at the sides, bottom, or top of the fuselage. Hatches are usually built into larger general aviation aircraft and commercial and military aircraft that carry passengers or cargo. Hatches generally have controls allowing them to be operated from inside or outside of an airplane by using quick-opening compression devices.

Hydrazine A fuel used in the emergency power unit on aircraft such as the F-16. This substance is extremely toxic and caustic.

Jetliner This term commonly describes jet-powered cargo or passenger-carrying aircraft.

Landing gear The understructure—usually, wheels, tires, and struts—supporting the weight of an aircraft when it is not in the sky. Also called the *undercarriage*.

Leading/trailing edge The forward/rear edge of an airfoil. Applies to tail surfaces, wings, propeller blades, and so on.

Longerons The principal longitudinal (lengthwise) structural members of the fuselage.

Microjets This term refers to a class of aircraft called very light jets (VLJs).

Military aircraft These may be any kind of airplane owned and operated by military forces.

Occupants Passengers and aircrew on board an aircraft. Also referred to as "souls on board" (SOB).

Ordnance This term refers to ammunition, bombs, rockets, or other explosive materials that may be carried on military aircraft.

Ribs A part of the skeletal structure of an aircraft wing that gives the airplane form, strength, and shape.

Rudder The upright, movable part of the tail assembly, which controls the direction of the aircraft.

Short take off and landing (STOL) This class of aircraft is designed to take off or land on short runways.

Skin The outer covering of an aircraft, including the fuselage and wings.

Slats Movable auxiliary airfoils whose primary function is to increase the aircraft's stability. This term usually references the fixed horizontal tail surface of the aircraft.

Spars The principal structural members, or beams, of a wing.

Specialized aircraft These are used for duties not found in mainstream commercial or military aviation, such as weather data gathering, drug interdiction, agricultural spraying and seed planting, law enforcement, medical evacuation, aerial firefighting, and other tasks.

Stabilizers These components are like fins, and they house the rudder that controls back-and forth movements (called *yaw*).

Stringer A long, heavy horizontal timber used for any of several connective or supportive purposes. (See *fuselage*.)

Tail This portion of an aircraft consists of vertical and horizontal stabilizers, rudders, and elevators. (See *empennage*.)

Trailing edge The rear edge of an airfoil. Applies to tail surfaces, wings, propeller blades, and so on.

Transport aircraft This term refers to aircraft designated for the purpose of carrying passengers or cargo (freight).

Ultralight aircraft This class of aircraft is characterized by a small airframe and motor. These machines are designed for recreational purposes and do not require an FAA registration number, airworthiness certificate, or pilot certification. These aircraft resemble hang gliders, and most have a single-seat configuration. Some ultralight aircraft, however, are designed for a pilot and one passenger.

Very light jets (VLJ) Very light jets (sometimes called microjets) are aircraft that use low-noise turbofan jet engines.

Wing root The point at which an aircraft wing is joined to the fuselage.

Wing strut This structure looks like a rod or pole and is connected to the airplane between the bottom of the wing and the bottom-side area of the fuselage. The strut provides additional strength for the aircraft.

REVIEW QUESTIONS

1. What is the outermost surface of an airplane called?
2. What are VLJ aircraft?
3. What term commonly refers to weapons, such as ammunition, bombs, explosives, or rockets that are carried on combat aircraft?
4. What are the side windows on passenger and private airplanes usually made of?
5. Why is it extremely difficult, if not impossible, to cut through the forward windshields on high-performance aircraft, such as military fighters or other high-speed airplanes?
6. Where is the empennage located on a fixed-wing airplane?
7. A wing strut connects the wing to the fuselage. What is the purpose of a wing strut?
8. What portion of a fixed wing airplane will passengers and/or cargo be located?
9. What does the term *general aviation* aircraft refer to?
10. All very large (heavy), large, and medium, and many small (light) aircraft are equipped with hydraulic systems. What are three components that make up these systems?
11. A circular partition that separates one aircraft compartment from another while giving shape to the aircraft is called what?
12. In which direction do plug-style hatch doors or covers open?
13. It is safe to assume that large-frame aircraft carry large amounts of cargo and fuel. In a post-crash fire involving such aircraft, what classes of fires might you confront?
14. You are responding to a successful emergency landing on a county road. The landing was necessitated by a bird strike through the windshield, and the pilot and passenger have facial injuries. The plane is a Cessna 172 general aviation aircraft. What part of this airplane must you and other people stay away from?
15. What is an APU? Does it have fuel lines? Does it emit hot exhaust gases that can burn you?

STUDENT EXERCISES

1. Using the aircraft illustrated in Figure 2-33, do the following:

   Circle the cabin (passenger seating) area of the aircraft.

   Circle the propeller danger zones.

2. Using the locations in the following list, write the appropriate letter on the aircraft in **Figure 2–33.**

   A. Most likely fuel tank locations

   B. Fuselage

   C. Empennage

   D. Nose

   E. Entry/Exit door location.

   F. Landing gear (3)

   G. Engine nacelles

   H. Emergency escape hatches

3. In your own words, describe the main differences between military combat aircraft and commercial passenger aircraft.
4. The twin-engine planes in Figures 2-33 and **2-34** share common design features with many larger aircraft. List some of these features.

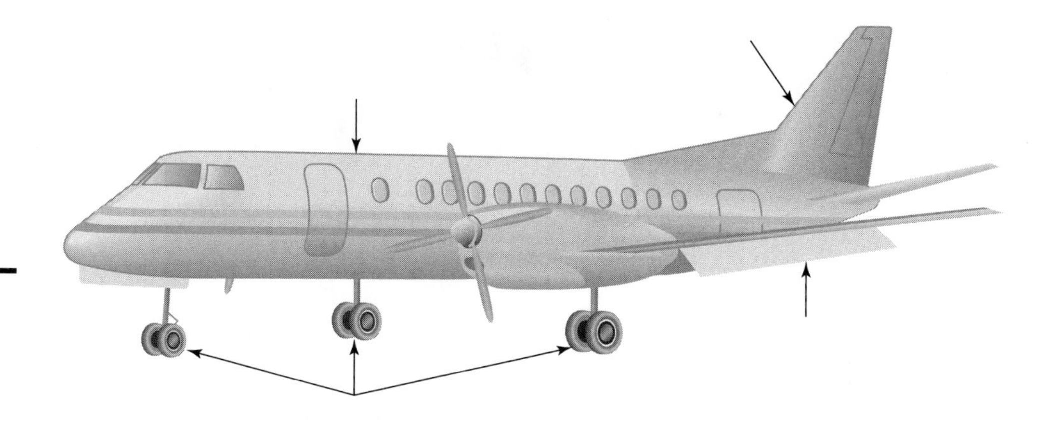

Figure 2-33 *A Saab 340 regional airliner. (Courtesy of Dave Armour.)*

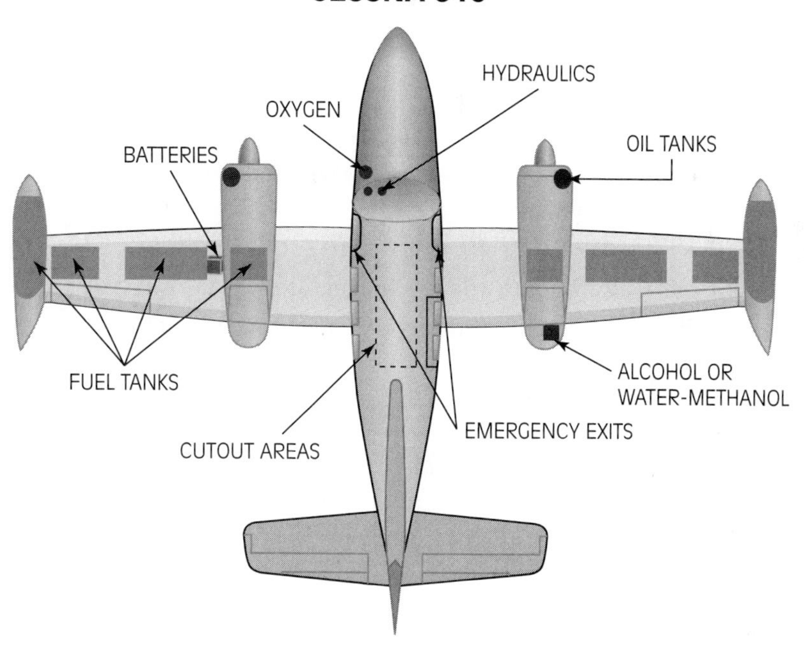

Figure 2-34 *A diagram of a Cessna 310. (Courtesy of Lahey Applications and Design Services.)*

5. Match each sentence description with the term in the left column that it best describes.

Egress System	A. Moves the aircraft. It can be a piston, jet turbine, or turboprop.
Engine	B. Provides minimal isolation from engine noise and engine heat
Nacelle	C. Like the skeletal system of a typical aircraft
Firewall	D. Refers to a means of quick escape from an aircraft
Framework	E. The housing that contains and covers an aircraft engine

AIRCRAFT HAZARDS

Learning Objectives

Upon completion of this chapter, you should be able to:

- Understand primary aircraft hazards, including fuel loads and hydraulic systems.
- Have a basic understanding of secondary aircraft hazards, including:
  - electrical systems
  - additional secondary hazards
  - aircraft construction materials
  - advanced aircraft composite materials
  - cargo/baggage/contents
- Understand military aircraft hazards.
- Understand specialized aircraft hazards, including:
  - liquid oxygen
  - radar
  - protruding devices
  - wing surfaces
  - specialized military aircraft hazards
- Understand personal protective equipment (PPE) for aircraft incidents.
- Understand how to use your issued PPE.
- Understand NFPA regulations for PPE.

INTRODUCTION

Aircraft have evolved tremendously since the Wright brothers took to the skies during the first years of the twentieth century. Modern aircraft are much larger than their predecessors, and they carry more fuel, passengers, and cargo. They are built from new construction materials that pose different concerns for firefighters than the materials that rescuers confronted a few decades ago. Fuels, hydraulic and electrical systems, composite materials, and the nature of payloads have changed the dynamics of how rescuers approach and manage aircraft accidents. Once you understand the probable reactions of these systems and materials within the context of an accident, the safer and more efficient your operations can become.

Airplanes are as diverse as automobiles and trucks. All aircraft share basic hazards, but different designs and uses result in unique hazards. The amount of people, cargo, and fuel aircraft carry differs based on mission and on size, shape, and other design variables.

life safety
the protection of human life

Your highest priority in all aircraft incidents is **life safety**, a term referring to the protection of human life: the lives of rescuers, bystanders, and accident survivors.

PRIMARY AIRCRAFT HAZARDS

All first-responding rescue personnel must have a basic understanding of the kinds of aircraft they may have to deal. Most of today's aircraft (except for the most basic types, such as ultralights) possess several primary hazards. First, we discuss fuel load and hydraulic systems.

Jet A fuel
kerosene-type fuel with a flash point of 110° F to 115° F

Jet B fuel
kerosene-type fuel with a flash point of −16° F to 30° F

AVGAS
high-octane gasoline with a flash point of −36° F

Fuel Load

All aircraft contain fuel, which is either **Jet A**, **Jet B**, or **AVGAS** (aviation fuel). If you cannot remember statistics about the various fuels' specific ignition temperatures (or *flash points*), remember this: Jet fuel (Jet A and B) is in the kerosene family. (Some agencies, such as the military, use JP-4, JP-5, and JP-8 grades of fuel.) It burns similarly to and is extinguished in the same manner as kerosene, diesel, or gasoline. AVGAS burns similarly to the gasoline used in automobiles. Fuel quantities on aircraft are based on the size of the airplane and can range from hundreds to thousands of gallons.

Jet A has been the standard jet fuel type since the 1950s. It has a auto ignition temperature of more than 425° F. When in storage or transport, it may be identified by the United Nations ID number 1863.

Jet B is a lighter composition than Jet A, which makes it more dangerous to handle. Thus, its use is restricted to areas where its cold weather characteristics are absolutely required.

AVGAS is a high-octane fuel with a very low flash point to improve its ignition characteristics. Fires involving these fuels can be contained and extinguished using a Class B firefighting foam. The most common Class B firefighting foam is **aqueous film forming foam (AFFF)**.

aqueous film forming foam (AFFF)
fire-extinguishing agent that contains fluorocarbon surfactants and spreads a protective blanket of foam that extinguishes liquid hydrocarbon fuel fires by forming a self-sealing barrier between the fire and fire-sustaining oxygen

At a crash scene, all fuels may atomize (become mist), becoming readily ignitable. From a fire protection standpoint, your safety concerns and fire suppression tactics at an aircraft crash scene are the same for any type of hydrocarbon fuel.

Hydraulic Systems

Figure 3-1 shows an aircraft hydraulic system found on this typical jetliner, which resembles the BAE 146/Avro RJ airliner used worldwide for passenger and cargo service. Notice that for safety, this aircraft has four hydraulic systems, which allow "backup" hydraulics in the event of a hydraulic system failure.

Small, low-speed aircraft use cables and pulleys to control the movement of aircraft in flight. Even these airplanes, however, may have a retractable landing gear that uses some form of a hydraulic system. Larger, faster airplanes use hydraulic

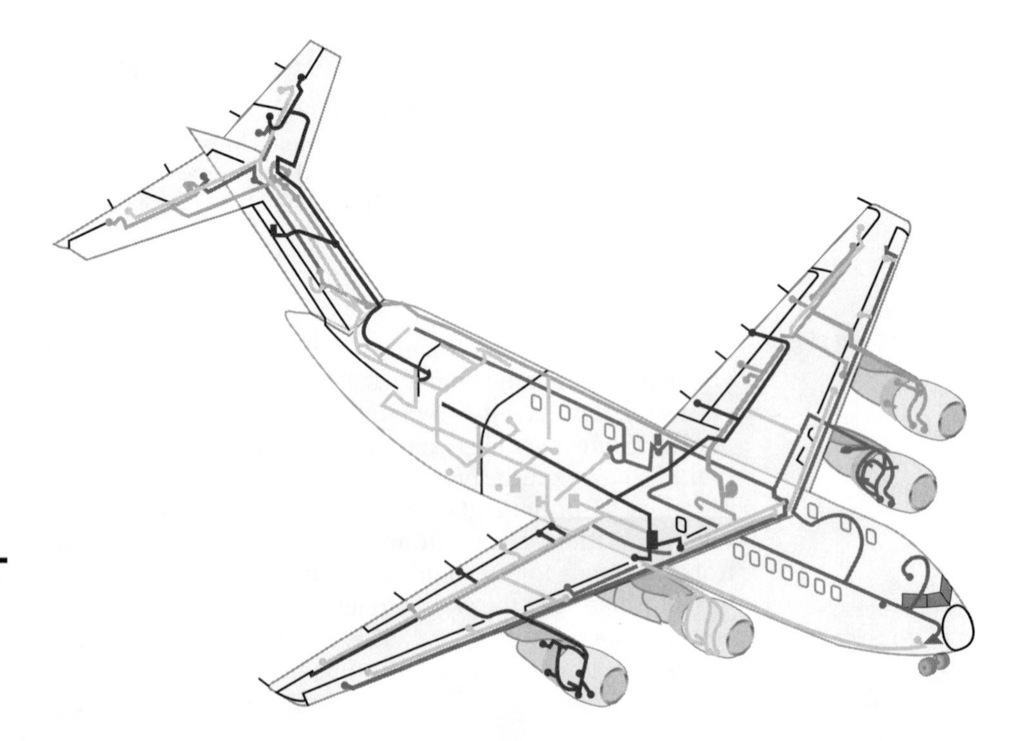

Figure 3-1 *A typical large-aircraft hydraulic system. (Courtesy of the U.S. Air Force.)*

systems as the primary system that enables the aircraft to operate while in flight. High-pressure hydraulic lines run throughout these aircraft to control vital systems, such as steering, brakes, and the ability to control the aircraft's pitch (nose up or nose down). Pitch, along with engine throttle settings, cause the aircraft to gain or lose altitude.

• Caution
Aircraft hydraulic systems are under high pressure: some systems hold hydraulic fluid with as much as 5,000 pounds of pressure within their pumps and tubing. Even if the engines of a crashed aircraft are not operating, cutting through tubing containing hydraulic fluid under such great pressure can cause serious injury.

Aircraft hydraulic systems are under high pressure: some systems hold hydraulic fluid with as much as 5,000 pounds of pressure within their pumps and tubing. Even if the engines of a crashed aircraft are not operating, cutting through tubing containing hydraulic fluid under such great pressure can cause serious injury.

Skydrol LD-4 and Skydrol 5 hydraulic fluids are fire-resistant under normal conditions, but may become unstable during an accident. Skydrol is a clear, purple, oily, and odorless liquid. It is a phosphate ester-based fluid blended with performance additives.

Skydrol is considered to be environmentally friendly and noncarcinogenic (cancer causing). Data on exposure to hydraulic fluids (based on less than one mouthful being swallowed), however, indicate that they are slightly toxic. Signs and symptoms of overexposure to hydraulic fluids include irritation, nausea, headache, vomiting, blurred vision, tearing of the eyes, redness, and defatting and cracking of the skin. Organs particularly vulnerable to repeated or prolonged exposure are the eyes, skin, and respiratory and gastrointestinal tracts. Exposure of the eyes to this substance may cause severe pain. Skin exposure may cause drying and cracking, which if unattended could result in complications such as dermatitis or secondary bacterial infection. Inhaling vapors or mists of Skydrol-type fluids may cause upper-respiratory tract irritation, including irritation of the nose and throat, and even tracheitis or bronchitis. People with asthma may be especially vulnerable to exposure to these liquids. Tributyl phosphate, an ingredient of Skydrol, may cause nausea and headache if inhaled or ingested and may be very harmful if ingested. Chronic exposure to Skydrol may cause urinary bladder damage (based on animal studies).

If hydraulic fluid enters the eyes, immediately flush with water for at least 15 minutes. Seek medical attention as soon as possible. In the event of skin contact, immediately remove contaminated clothing and flush the skin with water. Exposure of large areas of skin requires immediate medical attention. All contaminated clothing should be bagged and isolated, so that after the emergency, it can be laundered or disposed of, following local health and environmental regulations.

Take precautions when fighting fires involving this fluid: wear full firefighting clothing and use a self-contained breathing apparatus (SCBA).

boiling liquid expanding vapor explosion (BLEVE) often called a "blevie;" when a closed container of liquid is exposed to excessive heat or flame impingement, the catastrophic container failure that often results in fragments of the container dispersing with violent force

Most hydraulic fires result from pressurized hydraulic fluid exiting a pinhole leak and then coming in contact with an ignition source. Furthermore, hydraulic reservoirs, lines, and containers may "**BLEVE**" when exposed to excessive heat or flame impingement. A BLEVE is a **boiling liquid expanding vapor explosion**. This happens whenever any amount of liquid inside a container is heated to the boiling point and vaporizes. As the vapor is further heated, it expands, rupturing the container.

Jet engine turbine oil also contains a phosphate: tricresyl phosphate (TCP), listed by the military as MIL-L-23699E. The following is taken directly from TO 00-105E-9 *Aerospace Emergency Rescue and Mishap Emergency Response Information*:

> It is a synthetic hydrocarbon and is a lubricating oil additive. This liquid is brown in color with a mild odor. This product is determined to be hazardous, but is not expected to produce neurotoxic effects under normal conditions of use and with appropriate personal hygiene practices. Avoid inhaling mists. Overexposure to TCP by swallowing, prolonged or repeated breathing of oil mist, or prolonged or repeated skin contact may produce nervous system disorders including gastrointestinal disturbances, numbness, muscular cramps, weakness and paralysis. Paralysis may be delayed. Wash any exposed skin thoroughly before eating or smoking. Keep away from feed or food products. Do not use on food processing machinery. Store in a cool, dry, well-ventilated area away from heat.

• Caution
This product is determined to be hazardous, but is not expected to produce neurotoxic effects under normal conditions of use and with appropriate personal hygiene practices. Avoid inhaling mists.

SECONDARY AIRCRAFT HAZARDS

Secondary hazards do not, under any circumstance, eliminate or lessen the amount of injuries an unprepared responder may encounter. Secondary hazards must be taken seriously and always approached safely and with calculation.

Electrical Systems

Many aircraft have DC (direct current) as well as AC (alternating current) electrical systems. These include generators powered by auxiliary power units, magnetos, and RAM turbines. Electrical wires can produce shocks if cut. Large and medium-frame aircraft can contain miles of electrical wire that may pose entanglement and, if damaged, shock concerns for victims as well as rescuers.

• Caution
Large and medium-frame aircraft can contain miles of electrical wire that may pose entanglement and, if damaged, shock concerns for victims as well as rescuers.

Aircraft Batteries

Aircraft batteries, as shown in **Figure 3-2,** may be 12 or 24 volts. The electrical master switch and the auxiliary power switch must be shut off prior to disconnecting a battery. The locations of aircraft batteries, electrical master switches, and auxiliary power switches vary from aircraft to aircraft. You must be familiar with the locations of these items. Whenever possible, contact your local airport and schedule familiarization training on aircraft typically utilizing the airspace in your area.

! Warning When aircraft batteries are subjected to fire, they usually yield toxic and explosive vapors. Batteries may be alkaline, lead-acid, nickel-cadmium, or lithium thionyl chloride in type.

• Caution Overheated batteries should be removed to the outdoors if it is determined safe to do so. It is best to use water fog to lower the battery's temperature.

Overheated Aircraft Batteries When aircraft batteries are subjected to fire, they usually yield toxic and explosive vapors. Batteries may be alkaline, lead-acid, nickel-cadmium, or lithium thionyl chloride in type. An overheated battery is hot to the touch and yields a pungent smell as a result of venting fumes. It may also leak hazardous, corrosive chemicals, such as sulfuric acid or potassium hydroxide. Nickel-cadmium batteries contain an electrolyte liquid that is a strong, corrosive alkali. If you notice discoloration around the battery vent port, access panel door, or case, it indicates that the battery is overheated and venting corrosive, toxic gasses. Overheated batteries should be removed to the outdoors if it is determined safe to do so. It is best to use water fog to lower the battery's temperature.

A sound similar to popcorn popping indicates a battery is in **thermal runaway condition**, which releases toxic, corrosive gases. If this occurs, the aircrew should shut down all electrical loads from the battery and evacuate the proximity of the aircraft until the noise and venting ceases.

Open the battery compartment and assess the situation to deal with suspected overheating or thermal runaway, or to extinguish a fire. If conditions permit, you may elect to let an overheated battery cool off. This is accomplished more quickly if the battery compartment door is opened and the battery is evacuated to a safe distance from the aircraft (if no other concerns such as extrication of aircraft occupants or fire are present). To deal with a battery fire, use a fire-suppressing agent

Figure 3-2 *An aircraft battery.*

• Caution
A sound similar to popcorn popping indicates a battery is in thermal runaway condition, which releases toxic, corrosive gases.

• Caution
Cutting or disconnecting a battery cable while there is an electrical load may cause a spark, which could ignite fuel vapors.

• Caution
Emergency shutdown procedures vary according to the aircraft type and the circumstances encountered during a specific rescue operation.

thermal runaway condition
electrochemical reaction that causes a battery to overheat, release toxic vapors, spew electrolyte, and very likely explode

appropriate to the battery type. Halon and carbon dioxide are effective for suppressing fires in alkaline or nickel-cadmium batteries, while a Class D extinguisher is recommended for controlling fire in a lithium battery.

As always, when dealing with any components of an airplane's electrical system, you must know the locations of the electrical master switch and the auxiliary power switch and shut these off prior to disconnecting a battery. Keep in mind that although it is difficult to remember the battery locations on every type of aircraft, as you train yourself on the aircraft that frequent your area, you will become increasingly familiar with the various locations of batteries in airplanes. Additionally, you will learn that some aircraft have more than one battery and more than one battery location.

Remember: Cutting or disconnecting a battery cable while there is an electrical load may cause a spark, which could ignite fuel vapors.

And, as when dealing with hydraulic fluids, take precautions when fighting battery fires: wear full firefighting clothing, including protection against corrosive liquids and vapors, and use a self-contained breathing apparatus.

Emergency Shutdown Procedures

Emergency shutdown procedures vary according to the aircraft type and the circumstances encountered during a specific rescue operation. You should attempt an emergency shutdown *only* if you know and understand the procedure for the aircraft you are dealing with.

These procedures are methodical and often sequential:

First, as soon as safely possible, immobilize the aircraft by placing wheel chocks in front of and behind the wheels. When doing this, take into consideration the location of jet engines or propellers. An aircraft with multiple engines may be chocked at the nose wheel. If it is a jet with intake in the nose, approach the front landing wheel from slightly behind the nose, then chock the wheel. If the airplane has two or more wing-mounted engines, or has rear-mounted engines, chock the nose wheel. Train and discuss this technique with your nearest airport authorities or airport fire department.

The next step, when performing shutdown procedures from the cockpit—**cockpit shutdown procedures**—usually is to shut the aircraft throttle to the idle, then "off" position.

Then, pull the aircraft's "T" handle to arm the onboard fire suppression system and stop the hydraulic, electrical, and pneumatic systems for each engine. Most airplanes, except for small aircraft, are equipped with this emergency handle, which is located either by the throttles on the center console (the left cockpit console in fighter aircraft) or on the forward control panel or overhead panel. These devices are usually painted a bright color, to make them easy to spot. Note that some "T" handles are shaped like the letter "L," yet *might still be referred to as "T" handles.* (This is comparable to referring to all cola drinks as "Coke.") If an aircraft has an auxiliary power unit, the APU should have a "T" handle.

■ Note

If the aircraft electrical system has been shut down or disabled, the "T" handle and fuel transfer switches are likely to be inoperable. This may result in the continuous siphoning of fuel through a damaged or severed fuel line.

● Caution

Cutting a battery cable or using the quick disconnect while there is still an electrical power load may cause a spark, which could ignite spilled fuel or flammable vapors.

emergency shutdown procedures
these are methodical and often sequential steps for shutting down an airplane. They include shutting the aircraft throttle (or throttles) to the idle, then "off" position; turning off electrical systems; safetying ejection seats (if possible); turning off fuel selection switches, and so on

cockpit shutdown procedures
methodical shutting off of aircraft engines, fuel pump switches, electrical power switches, hydraulic systems, and so on

If the aircraft electrical system has been shut down or disabled, the "T" handle and fuel transfer switches are likely to be inoperable. This may result in the continuous siphoning of fuel through a damaged or severed fuel line.

After you have engaged the aircraft's "T" handle, press the button or switch beneath it that activates the release of fire-extinguishing agent.

Once you have turned off the airplane's throttle and engaged the "T" handle, you must turn off the electrical master switches and disconnecting the battery. (Some batteries are equipped with "quick disconnects," which allow the responder to disconnect batteries without having to cut cables or use wrenches.) *Remember:* Cutting a battery cable or using the quick disconnect while there is still an electrical power load may cause a spark, which could ignite spilled fuel or flammable vapors.

If a disabled aircraft has an ejection seat, the cockpit area is an especially dangerous place, because it contains an emergency ejection seat or jettison lever in addition to other emergency handles. In such a situations, do not operate a lever or switch unless you know what the result will be. Unless the device is a "T" or an "L" handle to arm a fire-suppression system, remember this rule of thumb: Any device that is painted yellow with black stripes or painted red has a critical function and/or involves a process that poses danger. Preplan to arm yourself with basic, generic information about these handles, so that you know the difference between a "T" handle and an emergency jettison or ejection handle.

Shutting Down Small General Aviation Aircraft According to the pilots consulted by the authors, the following is the most common advice regarding general aviation airplanes for the non–ARFF-rated responder:

- Several knobs may be located on the lower center of the instrument panel, depending on the complexity of the airplane. The *carburetor heat knob* has a square face: Ignore this knob. To the *right* of the carburetor heat knob is the *throttle,* which is *black.* The throttle has a friction lock. *Pull the throttle aft* (toward you) to reduce fuel flow. Note that even fully reduced fuel flow may not stop the engine(s). To the *right* of the throttle there may, or may not, be a propeller control: Ignore it.
- The fuel *mixture control* knob is to the *right* of the *propeller control* (if there is one) or to the *right* of the *throttle,* if no propeller control is present. The mixture control knob is *red,* has raised points around its circumference, and has a friction lock. *Pull the knob aft* (toward you) *to starve the engine.* This accomplishes a complete engine shutdown within a few seconds (depending on ambient conditions, throttle settings, and so on).
- Fuel selector systems are not relevant to this process: Ignore them. *The most important point is:* Shut down the engine by first reducing the throttle to idle, and then reducing the fuel mixture by pulling the mixture control knob to full aft until the engine stops.

Waiting to Execute Shutdown Procedures Whether or not you are familiar with emergency shutdown procedures for the aircraft in a given situation—but especially if you are not—it may be wiser to wait for better-trained personnel to arrive at the scene and execute the procedure. For instance, if an accident or emergency landing occurs close to an airport, it may be dealt with by airport firefighters. Or, once the scene is determined to be safe for entry, other key personnel such as aircraft mechanics, airport employees, or other qualified people who know the location and function of the aircraft's various emergency switches, can complete shutdown. (Also, these trained personnel will place pins in the landing gear struts of aircraft equipped with retractable landing gear.)

Try to obtain the estimated time of arrival of these trained personnel and act accordingly.

! Warning Do not operate a lever or switch unless you know what the result will be. Unless the device is a "T" or an "L" handle to arm a fire-suppression system, remember this rule of thumb: Any device that is painted yellow with black stripes or painted red has a critical function and/or involves a process that poses danger.

Additional Secondary Hazards

Other secondary aircraft hazards include dangers from severe tire and wheel wear, jet engine intake and propellers, and engine noise.

Danger from Tires and Wheels Airplane tires and wheels, which are made from aluminum or magnesium alloys, are subjected to some of the greatest stress experienced by aircraft components. This is a result of aircraft weight, repeated landings, slowing and stopping from fast landing speeds or takeoff aborts, use of the brakes (especially if a large airplane diverts to make an emergency landing on a short runway at small local airport), high tire pressures, "dragging brakes" (i.e., the pilot is riding the brakes, or the system malfunctions and the brakes do not fully disengage when they should), corrosion, and physical damage. These factors contribute to overheating, possible fire or, at the very least, *hot brakes*. This emergency condition can occur as a result of grease or other materials in the wheel assembly catching fire and weakening the wheel. This can escalate into a wheel assembly or wheel well fire. Some aircraft have sustained wheel or wheel-well fires that caused violent wheel disintegration in flight.

• Caution Shut down the engine by first reducing the throttle to idle, and then reducing the fuel mixture by pulling the mixture control knob to full aft until the engine stops.

Also, aircraft tires are usually equipped with fuse plugs that melt and "pop out" to relieve tire pressure and minimize the likelihood of catastrophic wheel failure/disintegration. Even if fire doesn't occur, wheel overheating may activate these plugs, which can be propelled with enough force to cause harm if they come in contact with a bystander.

If an aircraft wheel suffers catastrophic failure, magnesium wheel fragments can travel at explosive speed for 300 feet on either side of the wheel assembly. Personnel exposed to this situation may be killed or seriously injured by this shrapnel. In addition, it can puncture fuel tanks, fuel lines, hydraulic lines, and so on, resulting in spilled fuel, fire, or explosion.

! Danger If an aircraft wheel suffers catastrophic failure, magnesium wheel fragments can travel at explosive speed for 300 feet on either side of the wheel assembly.

Dangers from Jet Engine Intake and Propellers **Jet engine intake** is the forward opening on the engine where air is sucked in at high velocity. Twenty-five feet is the

Warning People have been ingested (sucked) into a jet engine even when standing 90 degrees to the side of, and at least 25-feet from, an intake.

Caution Jet engines' rear exhaust also is quite hazardous: It is hot, it may blow people off their feet, and it may generate blowing debris.

jet engine intake
the portion of a jet engine where air enters in at great velocity. The engine intake poses a potential hazard for responders

magneto
found in aircraft piston-type aircraft engines and generates electric current from a magnet that spins as the motor crankshaft operates

minimum safe distance from the front of an intake while engines are at idle. More distance is needed, depending on the aircraft, for larger jets or high-performance aircraft. People have been ingested (sucked) into a jet engine even when standing 90 degrees to the side of, and at least 25-feet from, an intake.

Jet engines' rear exhaust also is quite hazardous: It is hot, it may blow people off their feet, and it may generate blowing debris. The safety distances noted previously also apply to the jet engine exhaust. The safe distance from the rear of smaller jets can be as short as 100 feet, while the safe distance from the rear of larger jets and high-performance aircraft can exceed 280 feet. Similar to jet engine exhaust, APU exhaust is hot and may cause burns.

Like jet engines, propellers may cause injury or death to anyone within striking distance of them if they are turned in the normal direction of rotation and there is residual fuel in a cylinder. *Always* treat *any* propeller movement as if the magneto were turned on (which would supply power to the propeller).

The wind generated by propellers—called *propwash*—also can pose hazards.

An aircraft engine **magneto** is very much like the type in an early automobile engine or a lawn mower engine, because it is designed to create an electric current by rotating a magnet. This means that even if the electrical system has failed, when a piston aircraft engine's crankshaft turns, this current will "fire" the spark plugs. An aircraft engine can be legally certified only if it contains two sets of magnetos.

In the early days of aviation, before electric starters were installed in gasoline engines, airplanes were started by ensuring that fuel was getting to the engine, turning on the magnet, and clearing the area around the prop (except for a ground crewperson, who would turn the prop). This caused the engine crankshaft to turn the magneto, which would create a spark and start the engine.

Obviously, bumping into, or in any way manually turning, a propeller is potentially dangerous unless proper aircraft safety and ground operation procedures have been followed. As a rescuer, it is best to avoid the area around a propeller. If you are fighting an engine fire, ensure that no one activates any engine controls or attempts to manually turn the propeller.

Generally, when dealing with aircraft engines or propellers, you should keep Murphy's Law in mind at all times: If anything can go wrong, it will. In other words, regardless of how seemingly impossible or bizarre an occurrence may seem, *it can happen and probably has happened*. Always proceed with the utmost caution when approaching the engines or propellers of distressed aircraft.

In addition to the obvious dangers posed by airplane engines, engines generate noise that hinders vocal communications and causes trauma to your auditory senses that can result in hearing loss.

Aircraft Construction Materials

Today's aircraft designers create advanced construction materials for contemporary and future aircraft. This has resulted in a generation of airplanes that fly faster, higher, safer, and with better fuel efficiency.

Aircraft skin (its exterior material) has evolved from early metallic materials to thin, yet durable, aluminum to advanced composite materials. These materials also are being used for critical load-bearing structural components.

Each of the composite materials reacts differently to fire, impact, and general wear and tear. Some melt into a puddle, others fracture when overheated, still others delaminate or burn rapidly.

Construction materials for aircraft interiors are similar to those found in automobiles and also have evolved from ones previously used. The use of plastics has become common, and synthetics have replaced natural fibers. Such advancements produce a comfortable passenger environment that is more cost effective and durable than earlier versions. And, many aerospace and fire-protection engineers consider these advanced materials more fire safe under most circumstances than older construction materials.

The use of these newer materials, however, has affected fire-suppression tactics and concerns.

The combustion byproducts of some materials, depending on the circumstances, require special handling and disposal procedures. Burning rates vary, as do the toxic materials contained in the smoke generated by an aircraft fire. When fighting fires involving these materials, it is essential to wear personal protective equipment (PPE) and use a self-contained breathing apparatus in order to minimize injury. (Recommendations for PPE for managing a post-fire incident are detailed in Chapter 5.)

Advanced Aircraft Composite Materials

Composite materials are not new to aircraft construction. **Figure 3-3** shows an aircraft built of aluminum-tube framing and treated-cloth "skin," an example of an early form of composite material. Early aircraft were constructed of aluminum (as in the figure) or wood airframes, bulkheads, and wing structures. These were covered by fabric (usually canvas) treated with *dope,* which consisted of a mixture of tree sap, glues, or similar chemicals. The fabric was soaked in dope, which made it rigid, weatherproof, and durable.

Warning

As a rescuer, it is best to avoid the area around a propeller. If you are fighting an engine fire, ensure that no one activates any engine controls or attempts to manually turn the propeller.

During the 1940s, the British developed the foundation for modern aircraft materials. British scientists processed certain raw materials into carbon fibers, producing a strong, yet lightweight, material. The carbon-fiber threads made during this process are smaller in diameter than a human hair. These tiny threads are woven into larger-diameter threads, which in turn are woven into cloth. In a process similar to using premade fiberglass sheets, the cloth panels are layered, glued together with epoxies (called a *matrix,* which also can be glue, resin, or another cementing agent), and placed into molds. Other layering processes can be used, depending on the materials' manufacturer and intended use. Some methods require autoclaving (submitting materials to tremendous heat and pressure), while others employ a cold-lamination process.

Figure 3-3 *Treated-cloth airplane skin.*

Another form of composite material is pre-preg tape strips that are woven into patterns, not unlike weaving a basket, and then glued together using a matrix. Some composite threads or yarns undergo filament winding to make nose cones or pressure vessels. This process creates an extremely strong material.

The U.S. National Aeronautics and Space Administration (NASA) has been researching and developing these materials for many years. (Sometimes the materials are referred to as *man-made mineral fibers [MMMF]*; they sometimes are incorrectly referred to as *reinforcement* or *fiber*.) By using these remarkable composites, which are 25 to 30 percent lighter than aluminum, NASA is successfully achieving its goal to reduce aircraft weight by at least 25 percent in comparison to using aluminum and other conventional construction materials.

SpaceShipOne, designed by the Rutan Brothers, is an innovative example of how composite materials can be used to construct state-of-the-art air- and spacecraft. In 2004, this ship became the first private, manned spacecraft to reach the fringes of outer space (twice!), at an altitude of 367,442 feet. The all-composite spacecraft broke the previously set record attained by an X-15, which reached an unofficial world altitude record of 354,200 feet.

At the time of initial delivery to the commercial airlines as well as the U.S. Air Force, the cargo is composed of varying amounts of composite materials. After modifications during depot maintenance, this aircraft often contains a larger amount of composites replacing older style structural materials with more or different composite materials.

General aviation aircraft also are taking advantage of the low-weight and high-strength characteristics of these materials. The USAF C-37 aircraft (the military version of the Gulfstream V) pictured in **Figure 3-4** contains an assortment of composite materials. The Raytheon Premier 1 aircraft has an all-composite fuselage, as does the Cirrus SR 20. These airplanes have joined the ever-increasing ranks of general aviation (private), military, and commercial aircraft using these remarkable modern materials. Composites are now used in the construction of propellers, landing gear fairings, airframes, and exterior skin on an increasing number of airplanes. Newer generation aircraft are using advanced composites in critical areas such as the wing boxes, main wing spars, and pressure

AIRFRAME MATERIALS

NOTE:
Ailerons on A/c 521 & 542 are metalriveted sheet metal.

NOTE:
Composite materials are used extensively on this aircraft (Gulfstream V) to save weight and increase strength. Composite materials include metallic and nonmetallic structures for bulkheads, doors, flight controls, floor panels, fairing, necelies, panels, pylons, radome, tailcone, and winglets.

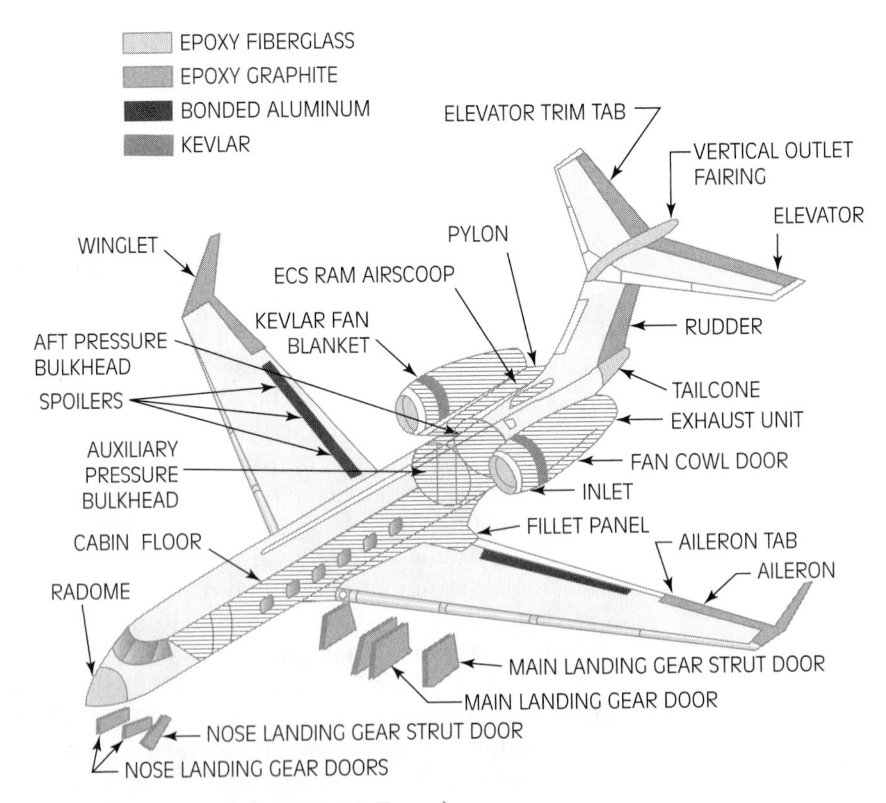

Figure 3-4 *Composites in a Gulfstream V. (Courtesy of the U.S. Air Force.)*

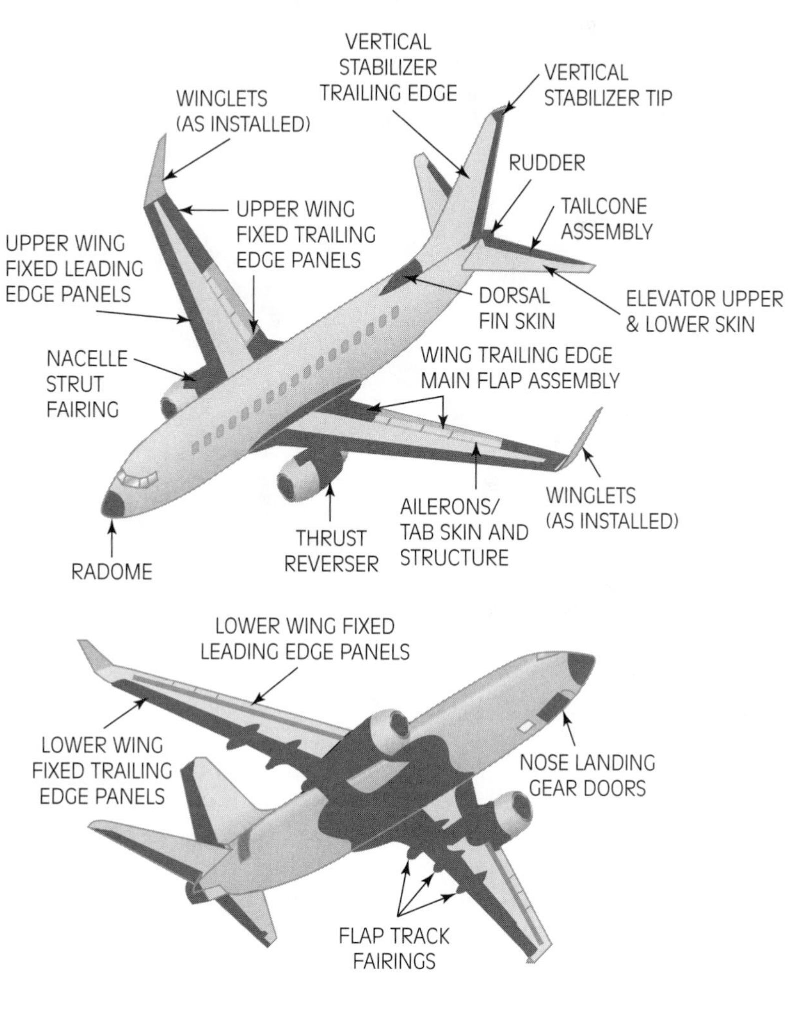

Figure 3-5 *The location of composite materials on Boeing 737-800. (Courtesy of the U.S. Air Force.)*

bulkheads. **Figure 3-5** shows an example of the amount of composites integrated into the construction of newer generation Boeing 737 aircraft.

Although composites are extremely strong materials—true carbon fibers have more tensile strength than steel—they are brittle and lack strength when subjected to *compression*. Forces of impact cause these materials to crack or shatter.

Figure 3-6 *Mojave Test composite-fiber burns conducted by the author, the U.S. Air Force, and other ARFF agencies. The black smoke is caused by burning jet fuel and by the burning, melting resin matrix.*

Fires Involving Advanced Composite Materials When subjected to the intense heat of pooled-fuel fires, as seen in **Figure 3-6,** the resin layers holding composites together are no longer able to bind the layers of cloth or threads. As the matrix melts and burns, it produces large amounts of black smoke (along with the smoke from burning hydrocarbon fuels).

As an uncontrolled "pooled fuel fire" engulfs carbon fiber composites, they delaminate and peel apart, resembling sheets of roofing tar paper from a distance. After delamination, the threads begin to separate and break into smaller lengths, like the whiskers left in a sink after someone shaves. At this point, separation occurs, and the fiber layers become very light, and the fire's heat column causes them to disperse. Glass fibers melt and form small beads; Kevlar® and carbon fibers decompose and disperse as the fibers become further detached from one another. When the *resins* from composite layers ignite, they themselves become a source of fuel for the fire. Some resins give off denser black smoke than others, but hydrocarbon fuel is usually the main source of the fire, in addition to smoke and soot release.

It can be difficult to detect the smolder phase of a fire involving composites, because often little or no *visible* smoke is produced. Smoldering epoxy is not easily influenced by wind currents, but yields large amounts of toxins and particulate.

• Caution
In all Class A fires, such as in structures or involving aircraft materials, the smoldering phase yields the most amount of toxins due to incomplete combustion.

In all Class A fires, such as in structures or involving aircraft materials, the smoldering phase yields the most amount of toxins due to incomplete combustion. Smoldering composites are dangerous because they contain more toxins than other materials, and a fire involving composites may progress undetected for a long time, while producing large volumes of toxic vapors. Carbon fibers burn *after* most or all of the matrix (resins) have burned away and the heat approaches 1,000° F.

When a fire's temperature reaches approximately 1,400° F, a red glow is visible. Smoldering epoxy doesn't generate enough heat to cause carbon fiber combustion, so, at this point, the carbon (also called carbon-graphite) fibers fibrillate. This means that they break down into microscopic hair-like pieces that are split, jagged, and deformed. Notice in **Figure 3-7** that the carbon-fiber composite's cloth weave is visible. Unidirectional carbon fiber tape engulfed in the flame for a certain period (depending on conditions) forms clusters, decomposed carbon fibers, and fiber ash. During a fire and the subsequent delaminating and breakdown, clusters form first.

Although fiber clusters are lightweight, they do not always remain floating, or suspended, in the air. Remember, however, that under certain conditions, airborne fibers can remain in a confined area, such a building or deep crater, for more

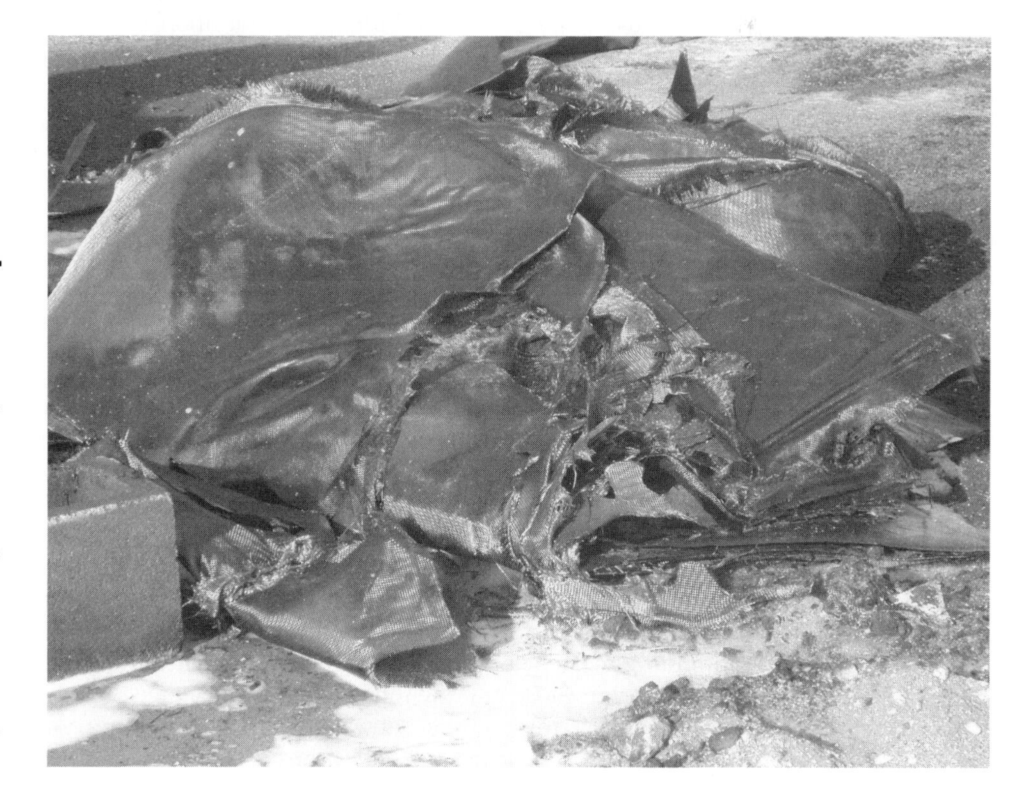

Figure 3-7 *Burned carbon-fiber composites. The burned carbon-fiber debris shown in this figure was rigid and hard to handle, and portions of it crumbled when attempts were made to grasp and pick it up. Such debris spreads additional ash and particulate at crash sites. Note that the cloth weave is visible.*

than 48 hours. Fibers and particulate may have settled on the ground or on other objects, then became airborne again when someone walked or drove through the debris field, or handled debris with fibers or particulate deposited on it.

Health and Other Hazards Posed by Advanced Composite Materials Although advanced aircraft composites have been used for many years, knowledge of their long-term and short-term health hazards is still limited.

Inhalation of particulates is the most serious potential health concern. In order to facilitate hazard assessment, fibers have been categorized as *high aspect ratio particles*. In other words, the smaller the particles are, the more damage they can cause. Toxic combustion products may adhere to the composites' fibers, which then carry the toxins into the respiratory tracts of any unprotected bystanders. Once inhaled, the fibers can become trapped deep in the alveoli of the lungs and cannot be efficiently expelled from the body. Alveoli are the small, elastic-like sacks, or balloons, where oxygen and carbon dioxide are exchanged during the breathing process. Normally, this process involves the depositing of oxygen and the removal of carbon dioxide and other waste products.

Once inside the body, the composite fibers can have variable and unpredictable toxic effects. Toxins may be transferred into the bloodstream, resulting in systemic poisoning (the specific effects are dependent on the nature of the toxins deposited by the inhaled fibers). Larger fibers can become trapped within the airways by impalement or electrostatic adhesion. In studies of natural and man-made fibers, a measurement of $L/D>3$ (a factor of fiber length and diameter) is a primary consideration for a substance to be classified as a hazard-posing chronic toxicity. What little human exposure data is available shows that these fibers, over time, eventually are expelled from the respiratory tract by the body's self-healing mechanisms.

Another potentially serious concern associated with the spread of carbon-graphite composite fibers is their tendency to become electrically conductive. These fibers can be easily attracted to unprotected electronic equipment and electromagnetic fields (as outlined in the U.S. Air Force *Aircraft Mishap Response Guidelines,* of October 28, 1993, and the *Aerospace Emergency Rescue and Mishap Response Information Manual 2006*). If subjected to sufficient number and concentration of these fibers, electronic equipment may be damaged or rendered inoperative.

Some testing has been done regarding this matter. Under controlled conditions, composites were burned near electronic equipment, but caused no problems associated with communications disruption or with emergency vehicles stalling due to electro-conductivity. Formal testing and post-fire analysis have concluded that downwind concentration and spread of composites in burning aircraft wreckage is far less than initially feared. Burn tests were conducted again in 2003, and these conclusions were verified.

However, researchers believe that electronic equipment still is potentially at risk when free carbon fibers are deposited in relatively large quantities on unprotected or closely proximal circuits and circuit elements. When carbon fibers settle on electrodes or circuits, they may cause computer circuit malfunction, arcs, or short circuits, which in turn, can blow 10–15 V fuses. Also, shorting transformers and arching voltage areas can cause equipment failure or malfunction.

■ Note
Currently, there are no published case histories of electrical power disruption or interference with motor vehicle ignition systems occurring as a result of an aircraft post-crash fire involving composites.

Some factories that manufacture composites sustain electrical fires on a regular basis. This is because after a certain manufacturing phase, small, hair-like threads are produced and float in the air currents within the building. These minute hairs collect on surfaces and objects like cobwebs (in spite of regular and thorough cleaning) and are attracted to electronic equipment, causing short circuits.

Currently, there are no published case histories of electrical power disruption or interference with motor vehicle ignition systems occurring as a result of an aircraft post-crash fire involving composites.

The Formal Research A report from the United States Environmental Protection Agency showed that when carbon-fiber/epoxy resin composite was incinerated at temperatures of 1,000° C, the epoxy resin was destroyed. The fibers were left intact with signs of thinning and pitting.

National Institute of Safety and Health (NIOSH) federal agency responsible for conducting research and making recommendations for the prevention of work-related injury and illness

A **National Institute of Safety and Health (NIOSH)** review conducted in 1980 showed that NASA performed a series of tests investigating the amount and type of airborne carbon fibers generated from burning carbon fiber composites and aircraft crash. The burn tests (simulated aircraft fires) indicated that heating the fibers reduced their diameter by partial oxidation and splitting (fibrillation). It was also concluded that a post-crash aircraft fire involving carbon-fiber composites would release fibers less than 3 microns in diameter and greater than 8 microns in length. Seventy-seven percent of the fibers were less than 1.7 microns in diameter. Most available test information has concluded that 24 percent of the fibers released in composite burns are in the respirable (breathable) range.

DOT/FAA/AR-98/34 "Health Hazards of Combustion Products from Aircraft Composite Materials" reflects the following data: Tests conducted at the Dugway Proving Grounds took direct samplings of fibers from the smoke plume of burning composites. Fiber concentration was also noted. All fibers with an L/D (lethal dose) ratio greater than 3 were counted with fiber concentrations determined during the 20-minute burning time. The results showed an average fiber concentration less than 0.14 fibers per cubic meter. This is ten times lower than the OSHA criteria for asbestos exposure. The opinion of some of the researchers quoted in this report is that carbon fibers should be treated in the same manner as fibrous glass (fiberglass) due to the "lack of toxicological data." Numerous test burns have been conducted by various government agencies, and laboratory data has been gathered. The original purpose of this test was to collect small fibers 200 to 400 feet downwind from a composite fire. Smaller fibers were detected, so additional sampling stations were placed downwind. This test, however, was not designed specifically for a post-aircraft-crash environment. Most people are not routinely exposed to burning composite materials on a regular basis. As of

■ Note
Most people are not routinely exposed to burning composite materials on a regular basis. As of this writing, there have been no documented deaths resulting from exposure to composites (burning or otherwise).

this writing, there have been *no* documented deaths resulting from exposure to composites (burning or otherwise).

! Warning **Treat aircraft mishaps involving burning composite materials as hazardous materials incidents until all pertinent data is known.**

Because data is still being obtained and information is still limited, it is best to play it safe. Treat aircraft mishaps involving burning composite materials as hazardous materials incidents until all pertinent data is known. All burned carbon-fiber composites are considered hazardous materials. They must be handled carefully and disposed of as hazardous waste. You can always downgrade your level of precautions as more is learned about the short- and long-term effects of exposure to burning composite materials.

The simplest means of preventing the spread of these fibers *is to keep the burned materials dampened* by using water or foam. Do not permit unnecessary personnel in the danger zone. All fire and rescue personnel must wear self-contained breathing apparatuses. Avoid contact with debris to prevent needle-stick injuries from stiff, jagged pieces of debris. If a fiber shard punctures skin, it must be removed carefully, because splinters will break off from the shard if it is removed incorrectly. Advanced crash recovery teams should follow special procedures to contain the burned debris, including spraying a fixant solution, consisting of one part polyacrylic floor wax to one part of water, on it. Study **Figure 3-8** and

Figure 3-8 *Crash recovery crews applying spray fixant to carbon-fiber composite debris.*

note that people make errors: The person on the right, who is applying fixant, has not applied tape to the tops of his boots.

Composite debris is characterized by what may look like clumps or strands of hair and delaminated sheets of what looks like sheet plastic or tar-paper. If this debris sites on dirt, it may be contained with agricultural soil tackifiers.

The Code of Federal Regulations 29 CFR 1910.120(q) covers all response, safety, planning, and training for hazardous materials emergency response operations. Many states have mirrored this regulation. (In California, for instance, the state regulation is Title 8, CCR 5192(q).) Check the applicable regulations within your state and find out which agency in your area is the ultimate authority having jurisdiction. Ensure that your agency regularly reviews its Standard Operating Guidelines (SOG) or Standard Operating Procedures (SOP) for hazardous materials incidents, composites mishaps, and aircraft accidents.

Cleanup of composite materials should be handled by professionals specialized in this field. The first responder should be concerned with extinguishment, mop-up, and initial containment of debris until wreckage disposal and composites management crews arrive on scene. Initial containment is simple and usually accomplished by keeping the debris damp with water or foam. Also, avoid spreading the materials to unaffected areas outside the crash site. Foam is the best choice until another spray fixant, such as poly-acrylic wax, becomes available at the scene. If the foam blanket evaporates or is washed away by rain or water, reapply it.

Cargo/Baggage/Contents

If you are a structural firefighter whose station is near an airport, you may be called on to assist in fighting a fire at an airport cargo support building or in a cargo aircraft. To prepare for these situations, it is wise to schedule frequent walk-through sessions of the airfreight facilities and cargo-carrying aircraft typical of the types that use local airspace.

Military cargo transports and some commercial cargo airlines carry freight on pallets, but the majority of civil cargo lines use containers called cargo cans. As seen in **Figure 3-9,** a cargo can is made of a thin, lightweight material and can be easily breached or cut open for firefighting access. Most of the cargo cans in use are made of thin Plexiglas, which allows responders to view the cans' contents without opening them.

Many cargo-hauling airlines ship hazardous cargo in cans bearing red-painted tops to provide quick identification of hazardous cargo. Your local airport fire department should have a list of emergency telephone numbers for cargo airlines. If there is no local airport fire department, then obtain these numbers from the *nearest* airport fire department.

As with any mode of transportation, such as maritime shipping, motor vehicle, and rail, the aviation transport industry is governed by regulations specific to air transport.

International Civil Aviation Organization (ICAO)
specialized agency of the United Nations that is responsible for developing international rules governing all areas of civil aviation

International Air Transport Association (IATA)
international industry trade group of airlines that helps airline companies in such areas as pricing uniformity and establishing regulations for the shipping of dangerous goods; publishes the all-important *IATA Dangerous Goods Regulations* manual

Title 49 CFR 175 states that "Shipping papers (such as those shown in **Figure 3-10**), or written notification of the presence of hazardous materials must be given to the pilot in command before the departure of the aircraft, and must be readily available to the pilot in command during flight. A copy of the shipping papers must accompany the shipment." These papers are carried in a special pouch; the location of this pouch may differ across airlines and aircraft.

Established in 1944 by the United Nations, the **International Civil Aviation Organization (ICAO)** is a specialized agency responsible for developing international rules governing all areas of civil aviation. Rules relating to transportation of commodities must comply with Title 49 CFR. The ICAO's safety responsibilities include a regulatory framework, enforcement and inspection procedures, and, when necessary, corrective measures regarding the airworthiness of aircraft, airport safety, personnel licensing, and international aviation rules. This agency's membership includes 183 contracting states (members).

ICAO works closely with the aviation community, the private sector, and other agencies and organizations, including the Universal Postal Union, World Health Organization, International Maritime Organization, **International Air Transport Association (IATA)**, Airports Council International, International Federation of Air Line Pilots' Associations, and International Council of Aircraft

Figure 3-9 *An aircraft cargo can. Note that the end walls are made of a thin, Plexiglas-like material.*

Owner and Pilot Associations. More information can be found on ICAO's Web site (http://www.icao.int).

Dangerous Goods *Dangerous goods,* often abbreviated as DG, are defined as substances capable of posing a significant risk to health, safety, or property when transported by air. They may be materials used routinely at work, in industry, or at home. Many common products are classified as regulated dangerous goods, including aerosols, paints, and assorted cleaning agents. All these materials can be dangerous and bear appropriate hazard labels, which include basic product hazard information.

Some people ship these products or carry them onboard commercial aircraft. Often this is unintentional, sometimes it is deliberate.

The International Air Transport Association (IATA) publishes regulations that govern the safe, efficient transportation of these materials and help identify and stop undeclared and other potentially hazardous shipments from being transported on aircraft. These regulations forbid many materials from being shipped on aircraft under any circumstances. The long list includes substances

Figure 3-10 *A dangerous goods envelope. These envelopes are designed to be highly visible, so that they can be spotted quickly by rescue personnel and aircrew.*

such as acetylene, ammonium nitrate, wet charcoal, ethyl nitrate, fulminate of mercury, and perchloric acid. Also forbidden are security-type attaché cases that contain lithium batteries. IATA presents these rules in language as clear, precise, and understandable as possible, so that they are comprehensible by as many people as possible.

The goal of creating laws requiring countries to adhere to IATA regulations was to produce a legal document that could easily be used in court by local authorities to enforce the rules. In addition, dangerous goods experts from the air transport industry meet twice a year, hosted by IATA, to review these regulations and ensure that they meet the industry's current safety needs. IATA also has a Dangerous Goods Training Task Force, which continually monitors the standards of dangerous goods training worldwide.

CASE STUDY

A courier unknowingly received a shipment of dangerous goods (an infectious substance affecting humans, the fungus "*Aspergillus*) shipped from Thailand through Tokyo to New York's JFK International Airport. According to the FAA, they placed the shipment in a yellow courier bag, labeled "HOLD AT AIRPORT FOR PICK UP." The courier company dishonestly declared the hazardous material shipment "Medical Shipment—Ambient Non-Hazardous, Non-Dangerous," contrary to the original and correct labeling. The shipment was transported on a U.S. passenger carrying flight from La Guardia Airport in New York to Dallas-Fort Worth Airport in Texas, and then on to San Antonio International Airport. An airline employee in San Antonio discovered that the shipment was a hazardous material and notified the FAA. The FAA alleged that the company also failed to provide and maintain emergency response information while the shipment was in transit.

This case highlights the importance of wearing personal protective equipment. Had this aircraft been involved in a crash or ground fire, responders would have been in proximity to this improperly shipped dangerous material.

Another violation of hazardous materials regulations involved commercial passenger-carrying aircraft transporting 50-pound containers of compressed hydrogen aboard two passenger aircraft in the United States. The labels on the containers indicated that the contents were "flammable" and "for transportation on cargo aircraft only." The airline transported the shipment on one of the passenger aircraft without providing written notification of the hazardous material to the pilot in command.

All people at cargo-carrying airlines who are involved in the movement of dangerous goods by air are required by law to have job-specific training, and this training must be provided or verified when an employer hires an individual who will be involved in the movement of dangerous goods. Refresher or recurrent training must be received within 24 months of previous training to ensure that employees' knowledge is current. Training for operators (anyone who packages or ships dangerous goods), or airline staff, must be approved by the competent national authority of the country where the operator or airline is based.

Packaging for dangerous goods must meet minimum quality standards. The package must be properly marked with labels that clearly identify the hazards contained in the package. Full documentation must accompany the shipment. Dangerous goods legislation is now in place worldwide for all modes of transport. Some cargo-carrying airlines carry all nine classifications of hazardous materials, with the exception of hazardous waste. The three main hazard classifications, however, are flammability, poison, and corrosive.

Proper documentation is a key element in safely transporting dangerous goods. In an emergency, it may not be safe to get close enough to a package to read its label or the name and address of the shipper, for example. This is why all dangerous goods shipments must be accompanied by a document that describes in detail the material being transported. For air transport, this is called a Shipper's Declaration of Dangerous Goods, which serves two purposes: It is a legal statement by the shipper that the package containing the dangerous goods has been properly prepared for transport, and it identifies the contents of all the dangerous goods in the shipment. (See the IATA Web site at http://www.iata.org.)

Handling labels identify packages containing dangerous goods and generally indicate the hazard inside the package. These labels are large (10 cm × 10 cm), so as to be readily visible and to convey that special care is needed when handling these packages. This is especially true when a package is damaged or leaking. Handling labels are also used to advise the handler of the need for any special care with the package. The simplest of these levels is the familiar "This Way Up" label, which has two side-by-side, black or red labels pointing upward.

Handling placards are similar to labels, but much larger. Placards are used on the side of transport units, such as trucks, containers, rail cars, and so on to indicate that these units contain dangerous goods. Examples of DG labeling for air transport are shown in **Figures 3-11** and **3-12.**

When you respond to an emergency involving a distressed aircraft, use whatever means available to determine whether the airplane is transporting known hazardous cargo and, if so, to identify that cargo. The pilot in command is required by law to have this data within his or her access. If the pilot is incapacitated or the aircraft has undergone a catastrophic accident, look for the specially marked pouch containing dangerous cargo information. Always assume HAZMAT is onboard and treat an aircraft mishap or incident accordingly.

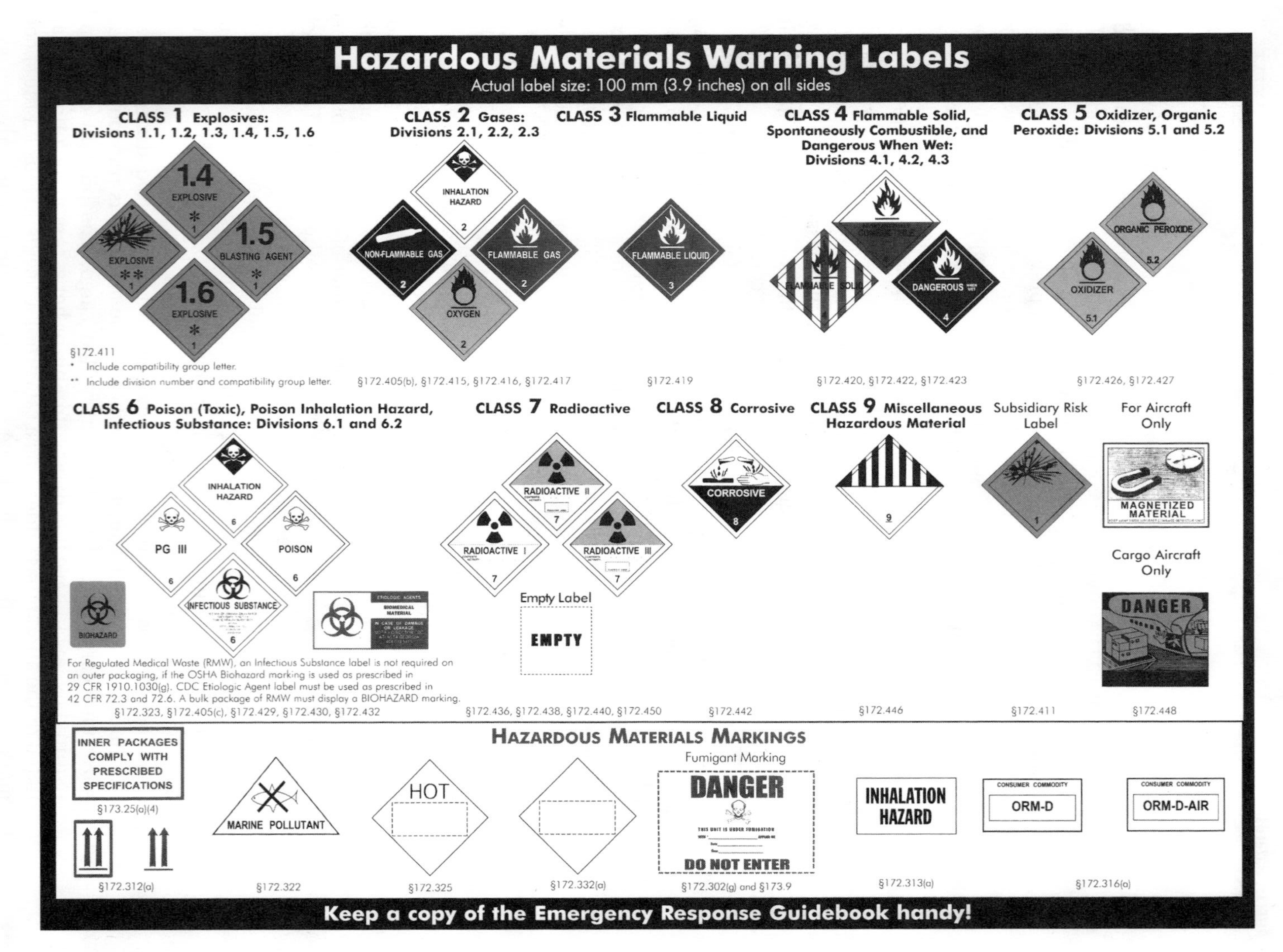

Figure 3-11 *Materials shipping labels.*

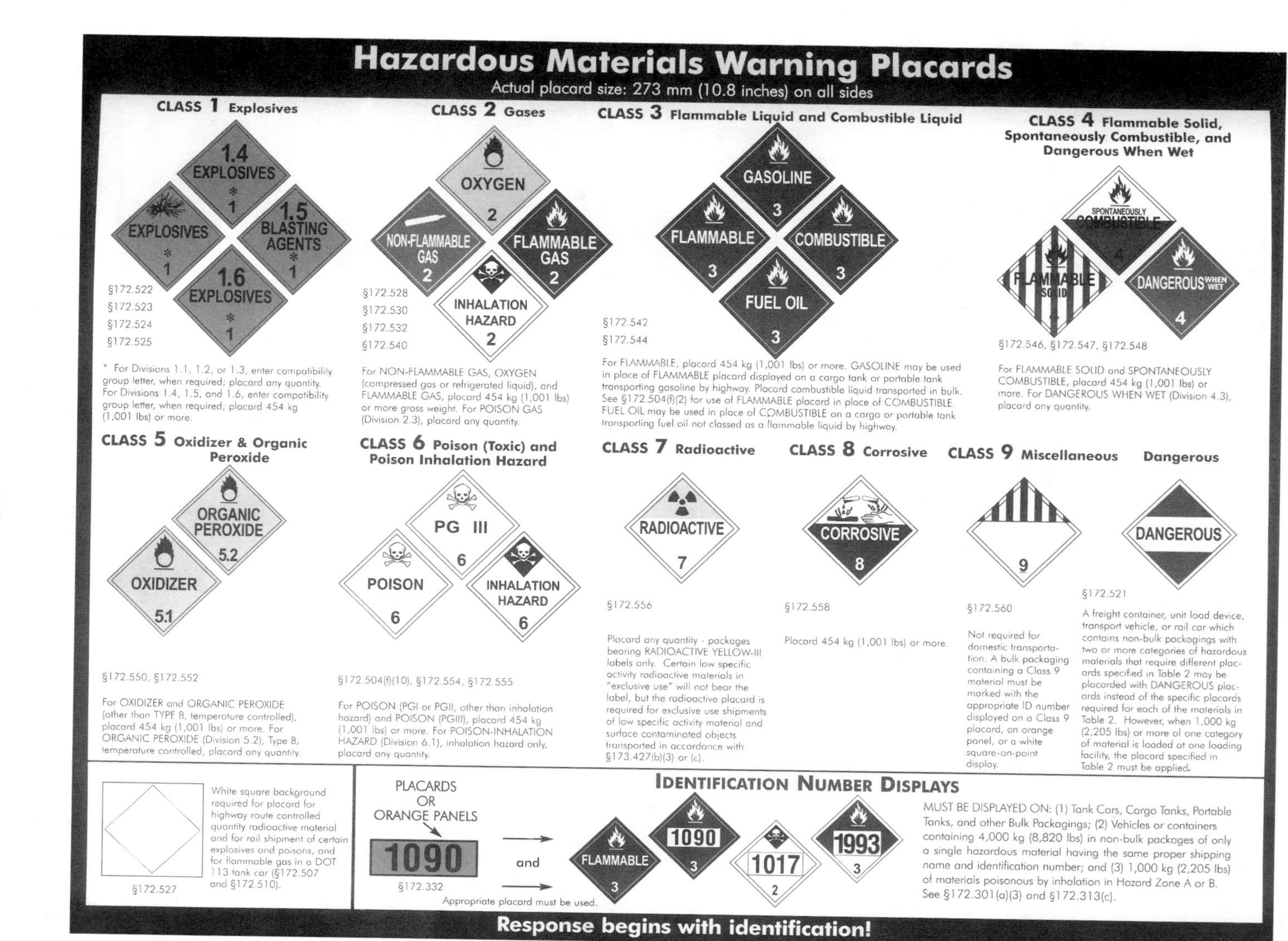

Figure 3-12 *Materials shipping labels.*

MILITARY AIRCRAFT HAZARDS

The military services use an assortment of specialized airplanes that include combat fighters and bombers, but also electronic warfare, communications, research, reconnaissance, and rescue aircraft. In addition, the military have cargo-/passenger-transport aircraft ranging in size from small frame to the giant C-5 transport. The Department of Defense also purchases and uses various general aviation (GA) and commercial aircraft, including the Cessna 377 and U-26 (Cessna Stationair); Boeing 707, 727, 737, 747, 757, DC-9, and DC 10; Learjets; and Gulfstream aircraft. These are used "as is" from the factory for a wide range of missions, including passenger transport, pilot training, weather data gathering, and aerial refueling. Some aircraft, such as the DC-10, have been modified to accomplish specialized mission requirements.

Should you be called to respond to an emergency involving military aircraft, knowing the type of aircraft you are dealing with better enables you to accomplish an initial hazard assessment on which to base your rescue actions. Handily, the alphanumeric designations of military airplanes supply information about the aircraft. The prefix letter categorizes military airplanes. For example, in the designation F-117, the letter *F* indicates that the aircraft is a *f*ighter type. The number *117* indicates which kind of fighter it is. In the designation C-130, the letter *C* indicates that the aircraft is a *c*argo transporter. The alphanumeric designation in F/A-22 Raptor indicates that the aircraft has multi-mission roles: The *F/A* prefix denotes that the airplane that is used as a *f*ighter and *a*ttack aircraft. Military aircraft also have names, such as the F-15 Eagle, C-130 Hercules, or C-17 Globemaster III. With this in mind, review the most commonly used aircraft prefixes.

Many American-built aircraft are used by other nations. Conversely, the United States uses the "Harrier" jet aircraft, which was designed in the United Kingdom. Following is a list of common prefix letters used in U.S. military aircraft.

A—Attack aircraft can be any size airframe, the same approximate size as a fighter or larger like the AC-130 gunship attack aircraft. The function of these airplanes is to provide air cover for endangered ground forces, and to conduct tactical operations against enemy ground targets using a variety of air-to-ground weapons. Examples include the A-10, AV-8, F/A-18, and F/A-22.

Figure 3-13 shows the A-10 aircraft. This airplane, nicknamed "Thunderbolt II" and "Warthog," has served its mission well by attacking ground targets. Ruggedly constructed, it is legendary for being able to take a pounding, yet return safely back to its base.

B—Bomber aircraft in the U.S. inventory are large-frame aircraft and include the B-1, B-2, and B-52. These aircraft drop bombs or deploy cruise missiles toward enemy targets.

The B-52 bomber in **Figure 3-14** can deliver a wide array of **ordnance**, which is a diverse menu of bombs, cruise missiles, and other offensive weapons.

ordnance
ammunition, bombs, rockets, or other explosive materials that may be carried on military aircraft

Figure 3-13 *An A-10 Thunderbolt II. The A-10 tank-busting aircraft (also called the "warthog") has proven itself as a rugged and powerful air-offensive weapon.*

Figure 3-14 *A B-52 Stratofortress. Old does not always mean inferior: This aircraft has been in flight for many years. In some Air Force families, three generations of pilots have flown this superior airplane.*

Figure 3-15 *A C-130. This versatile airplane serves many nations and performs many duties, including rescue, as shown.*

C—Cargo aircraft carry cargo or people, or a mixed payload of cargo and people. A few examples of cargo aircraft include the C-5, C-9 (the military version of the DC-9), C-17, C-40A (the military 737–700), C-141, and C130.

The venerable C-130 is used by many nations (see **Figure 3-15**) and performs cargo operations, troop transport/passenger flights, combat-attack, aerial refueling, special operations, rescue, weather data gathering, and firefighting duties with exemplary reliability.

Larger cargo aircraft include the C-5 (pictured in **Figure 3-16**), C-17, and C-141.

E—Electronic aircraft are used for surveillance, electronic countermeasures, or as airborne command posts and include the EA-6B and E-4 (Boeing 747B).

F—Fighter aircraft are designed for air-to-air combat, and include the F-14, F-15, F-16, F-18, F/A-22, and F-35.

H—Rescue aircraft, including the HC-130, are specifically configured to perform search-and-rescue operations. *H* also denotes helicopters. *HH* denotes military rescue helicopters.

Figure 3-16 *A C-5 Galaxy. This flying giant can carry mammoth payloads and massive numbers of people.*

K—Aerial refueling is performed by tanker planes, which are literally flying gasoline stations. They can deploy a boom (like a long extending pipe with wings at the tip). Examples of aerial refueling tankers include the KC-10 Extender (the military DC-10), shown in **Figure 3-17,** and the KC-135 (Boeing 707), seen in **Figure 3-18.** Modified aircraft for aerial refueling missions include the KC-130. The U.S. Navy currently uses the KA-4 and the S-3 Viking, which incorporate a refueling system called a **buddy store**.

buddy store
a modified, externally mounted fuel tank with the drogue-parachute type of refueling system

Figure 3-18 provides a close view of the refueling boom on a KC-135 Stratotanker. The pilot flying the aircraft needing fuel flies the aircraft close under the tail of the tanker plane. A boom operator extends then flys the refueling boom, guiding it to the first plane in coordination with the tanker pilot and the pilot of the airplane being refueled. The result of this complex aerial ballet is that the tail boom and refueling receptacle connect. Then, fuel is transferred from the tanker to the plane requiring refueling. Another method of fuel transfer is accomplished using special hoses that extend from the tanker. These hoses have drogue chutes, which resemble small parachutes. Just as with a refueling boom, the pilot of the aircraft being refueled connects the aircraft to a coupling on the end of the hose, then fuel can be transferred.

Figure 3-17 *A KC-10 "Extender." KC-10s carry liquid fuel cargo, standard solid cargo, and people.*

Figure 3-18 *A KC-135 "Stratotanker." Notice the refueling boom below this aircraft's tail—it is similar to the booms on KC-10 aircraft.*

M—Special Operations aircraft are assigned to specialized military forces that infiltrate behind military lines and perform classified missions. An example is the C-130, which has been converted to an MC-130. A variant of the RQ-1 Predator, called MQ-1, is capable of carrying and firing missiles.

N—Test bed aircraft may test navigation equipment or aircraft components, or perform specialized duties for aeronautical research and design. One example is the NKC-135A, which is a modified KC-135 that sprays water from its refueling boom onto a following aircraft to test the following aircraft's anti-icing equipment design or how that airplane is affected by icing conditions. The NKC-135E is a U.S. Army airborne electronic warfare laboratory.

O—Observation aircraft monitor ground activities, such as troop movements, or evaluate the accuracy of an air attack on a ground target. Observation aircraft are also used to assist in fighting wildland fires by providing data to firefighters on the ground and to water tanker pilots who drop retardant on the fires. Examples include the OV-1 and O-2. In the United States, OV-10 observation aircraft are used for wildland firefighting observation duties.

P—Patrol aircraft are use for patrol and detection missions. The Lockheed P-3 Orion, for example, is a military version of the Lockheed L-188 Electra passenger airplane. It is used by many nations for maritime patrol and submarine detection. It is also used in the U.S. for dropping fire retardant on wildland fires.

R—Reconnaissance aircraft gather intelligence data. Examples include the RC-135, the remotely piloted RQ-1 Predator, and the RQ-4 Global Hawk.

S—Strategic aircraft perform submarine detection and attack missions. An example is the U.S. Navy's S-3 Viking.

T—Training aircraft are used for various training duties, such as pilot or navigation training. Examples include the T-1, T-6, T-37, T-41 or larger aircraft like the T-43 navigation trainer (Boeing 737). The aircraft pictured in **Figure 3-19** is a T-6A Texan II, which is in use by the U.S. Air Force and U.S. Navy.

U—Utility aircraft (including helicopters) are used to shuttle small numbers of personnel by air travel. Examples of utility aircraft are the U-26 (Cessna Stationair), UC-12 (Beech 1900), and U-2.

V—VIP transport is a commercial airliner or business aircraft owned by the military. It may have custom luxury configurations or altered seating arrangements. Examples include the VC-9, VC-25, or VC-135.

The prefix *V* may also refer to vertical takeoff aircraft, such as the V-22 Osprey. The AV-8 Harrier is an attack aircraft with vertical takeoff and landing capabilities.

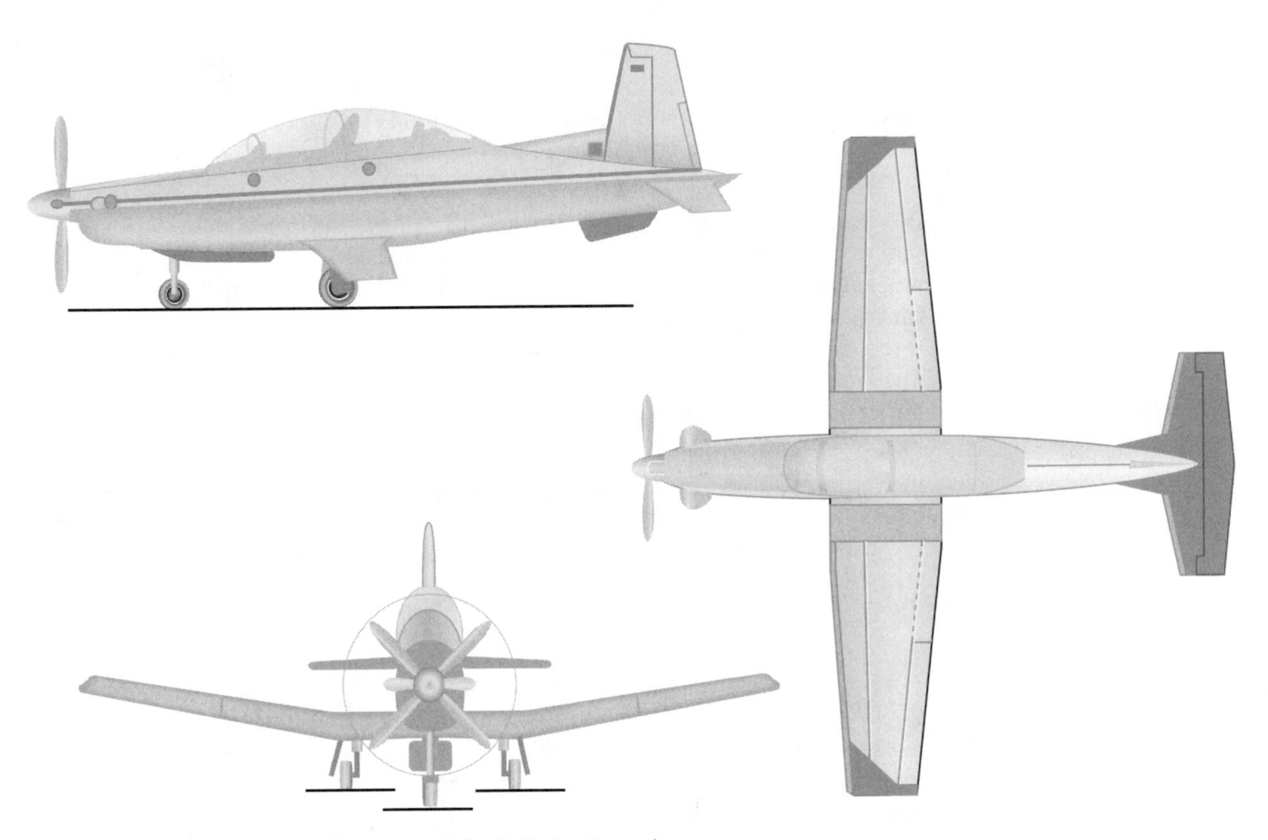

Figure 3-19 *A T-6A Trainer. (Courtesy of the U.S. Air Force.)*

W—Weather data gathering aircraft include the WC-130 and WC-135.

X—Experimental aircraft can be small-, medium-, or large-frame aircraft. These are currently in the developmental stages, such as the U.S. Navy's X-47.

Y—Prototype aircraft is a newer aircraft that is emerging for use. When the C-17 was in its evaluation phase, for instance, it was designated as the YC-17. Once the aircraft went into mass production/utilization, it became the C-17.

> **Danger** When HE has melted and resolidified, it is not only shock sensitive, but also heat sensitive. Such material may resemble a "mud pie." Do not step on or disturb it, as it may explode.

During military aircraft emergencies, responders must consider the possibility that the aircraft is carrying ammunition, bombs, explosives, or rockets (ordnance), which are carried on combat aircraft. Ordnance may still be attached to a downed military aircraft, or it may be scattered about the area. *Conventional* weapons and munitions are nonnuclear weapons and may be bombs, rockets, or missiles. (These may be carried as freight in a cargo aircraft.) Other common conventional weapons are machine guns and cannons. These can be either permanently affixed to the aircraft or carried in external weapons pods. Missiles

! Warning Always approach munitions at a 45-degree angle, whether they are still attached to the airplane or lying scattered on the ground at an accident site. If an aircraft is on fire, there is a chance that the heat may trigger weapons launch.

! Warning Flares present a critical fire hazard, because they burn extremely bright and with intense heat. Also, looking directly at a burning flare may cause eye damage.

• Caution LOX is extremely cold and can freeze human skin on contact, may ignite when in contact with other materials, can react with combustible and explosive mixtures, and may react when in contact with common Class A and B materials

and rockets are weapons that can be launched at enemy targets from the air, from the ground, or from ships at sea. They vary in warhead type, as well as explosive type. Some are designed to be incendiary (start fires), while others have explosives and are designed to target airplanes, motor vehicles, enemy personnel, tanks, or ships. They can be carried on wing tips, under the belly of an aircraft, or as internal weapons stores, or mounted on pylons beneath the wings. They also can be used in shoulder launchers by military personnel.

HE (pronounced "H-E") is an abbreviated term for high explosives, which are contained in bombs and missiles or cannon shells.

When HE has melted and resolidified, it is not only shock sensitive, but also heat sensitive. Such material may resemble a "mud pie." Do not step on or disturb it, as it may explode. Isolate the HE and inform explosive ordnance disposal experts about it.

Always approach munitions at a 45-degree angle, whether they are still attached to the airplane or lying scattered on the ground at an accident site. If an aircraft is on fire, there is a chance that the heat may trigger weapons launch.

Chaff and flares usually are mounted in the bottom tail section of an aircraft. A chaff dispenser is designed to eject large bunches of aluminum foil-like material in order to distract enemy weapons from successfully targeting an aircraft. Anyone close to a chaff dispenser can be seriously injured if struck by dispersing chaff. This also can happen when flares are dispensed. Flares present a critical fire hazard, because they burn extremely bright and with intense heat. Also, looking directly at a burning flare may cause eye damage. They usually are contained within affixed dispensers on aircraft, but they also may be carried aboard aircraft and manually thrown from aircraft during search-and-rescue missions. Flares are also likely to be contained inside survival kits and in flare pistols.

External fuel tanks are another hazard you may encounter during a military aircraft emergency. These tanks are teardrop shaped and carry varying amounts of fuel. As the term "external" implies, they are attached on the outside of an aircraft to provide added travel range. These tanks can break free from an aircraft during a crash, and rescuers sometimes mistake them for bombs. (There have been incidences of firefighting crews focusing their extinguishing efforts on a burning fuel tank, thinking it is a critical portion of the airplane.) These tanks may also be jettisoned (ejected away from the aircraft) after the airplane is on the ground.

SPECIALIZED AIRCRAFT HAZARDS

Specialized aircraft pose a wide range of potential hazards to rescue crews.

Liquid Oxygen

Some aircraft types, especially military aircraft and medevac helicopters, may carry liquid oxygen converters, such as those shown in **Figure 3-20. Liquid**

Danger When LOX spills on asphalt, it forms a potentially explosive shock-sensitive gel, which can detonate if it is driven over or stepped on. LOX cylinders may rupture in a fire.

oxygen, also called LOX, is oxygen that has undergone a cryogenic process that freezes the gas and compresses it so that it can be stored in a smaller tank or reservoir.

LOX is extremely cold and can freeze human skin on contact. It may ignite when exposed other materials. Also, it can react with combustible and explosive mixtures and may react when in contact with common Class A and B materials, such as wood, cloth, paper, oil, or kerosene. As LOX vaporizes, its gas is heavier than air and collects and remains in low-lying areas. When LOX spills on asphalt, it forms a potentially explosive shock-sensitive gel, which can detonate if it is driven over or stepped on. LOX cylinders may rupture in a fire.

Once released from its tank or reservoir, LOX expands in volume. Its expansion ratio is 860 times by volume when it is allowed to vaporize (i.e., return to

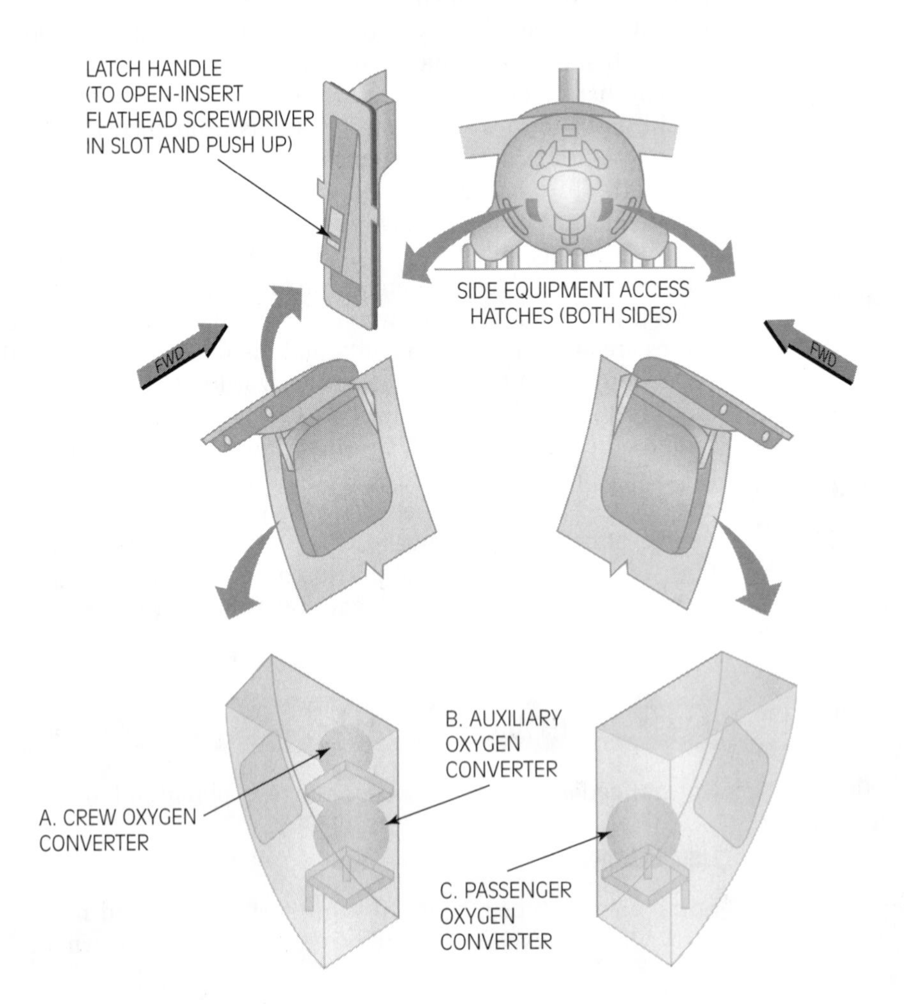

Figure 3-20 *Liquid oxygen converters. This illustration shows the locations of three converters on a C-17 cargo aircraft. (Courtesy of the U.S. Air Force.)*

Warning

PPE must be worn in the proximity of liquid oxygen or its vapors. Any turnout gear exposed to LOX vapors can easily ignite, because the pure oxygen saturates fabric. Any exposed clothing must be aired out before being used again.

gas form). You can place this information into perspective by trying this scenario: Set a coffee can on a desk or table, and pretend that it is an LOX container. Now, imagine 860 identical cans in a huge pile. That's a lot of oxygen, and it can sustain a tremendous fire.

Anything that has been exposed to LOX vapors or liquid burns more vigorously than it would otherwise. PPE must be worn in the proximity of liquid oxygen or its vapors. Any turnout gear exposed to LOX vapors can easily ignite, because the pure oxygen saturates fabric. Any exposed clothing must be aired out before being used again.

Gaseous oxygen is pure oxygen in its vapor (usual) form. It has many uses, such as medical treatment, furnishing pure oxygen in flight, and industrial uses. Figures 3-20 and 3-21 show oxygen stored in cylinders mounted in designated locations in general aviation, commercial, and military aircraft. (The cylinder in **Figure 3-21** contains gaseous oxygen and is stored beneath the right forward seat of this general aviation airplane. If the cylinder were involved in a fire, the results could be disastrous.) Some commercial aircraft types carry emergency oxygen generators, which are used in the event of a loss of cabin pressurization. These generators function as the result of a chemical reaction and pose a flammability hazard.

HE
an abbreviated term for high explosives, which are contained in ammunition, bombs, and missiles or cannon shells; pronounced "H-E"

Figure 3-21 *A pressurized oxygen cylinder. The cylinder in this figure is mounted below the front edge of the right-forward seat of this private aircraft.*

CASE STUDY

The crash of a Valuejet DC-9 passenger jet into the Florida Everglades claimed 100 lives. The cause of this notorious tragedy was oxygen generators in the lower baggage hold that were improperly labeled and illegally transported. These somehow became activated and started a rapidly burning fire, which quickly burned through the aircraft control cables making flight control impossible. Despite the heroic efforts of the pilot and copilot, no one on that plane stood a chance.

! Danger Any pressure vessel may explode when subjected to fire or severe impact forces, posing a fragmentation hazard. Whenever a pressurized cylinder (including a fire extinguisher) is heated to the point at which it loses its strength, there is danger of an explosion.

Any pressure vessel may explode when subjected to fire or severe impact forces, posing a fragmentation hazard. Whenever a pressurized cylinder (including a fire extinguisher) is heated to the point at which it loses its strength, there is danger of an explosion. **Figure 3-22** shows a complete walk-around breathing cylinder. On the opposite side of the partition in this airplane is a pressurized fire extinguisher. Some aircraft also carry nitrogen pressure cylinders and onboard fire-extinguishing systems containing pressurized Halon.

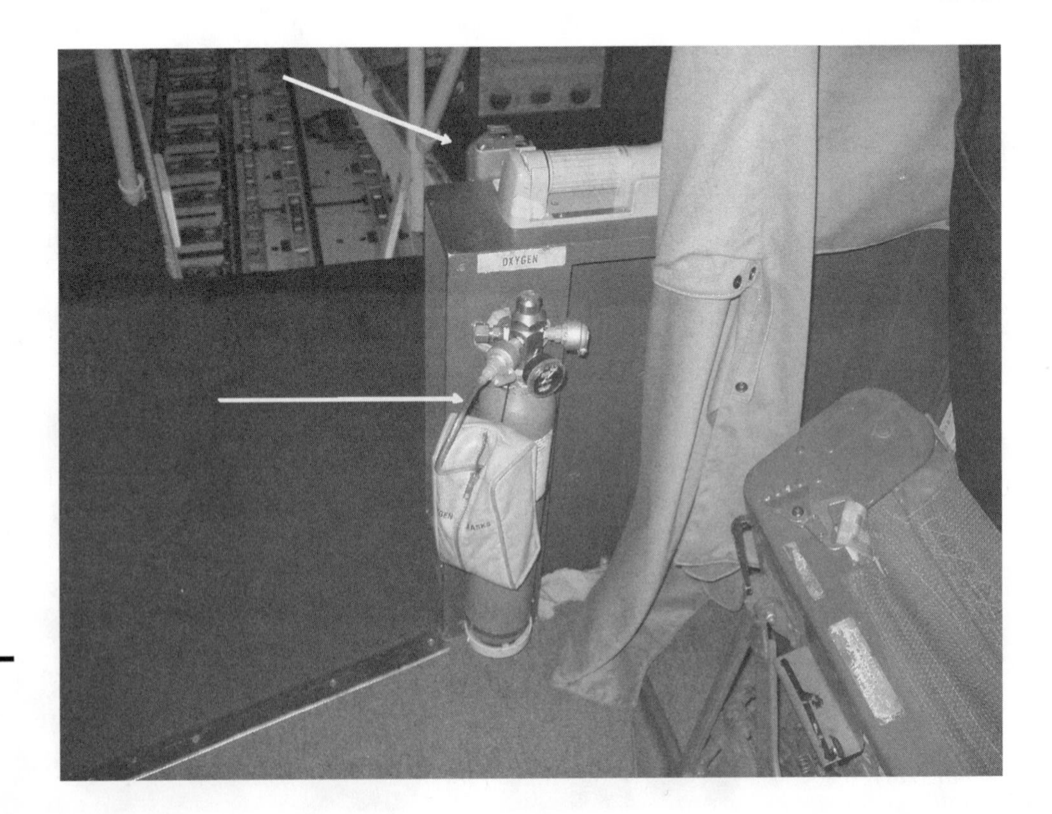

Figure 3-22 *A portable ("walk-around") air cylinder.*

● Caution

Aircraft radar can cause radiation burns, often called radar burn.

Radar

Radar systems also can be hazardous to rescuers. Aircraft radar can cause radiation burns, often called *radar burn*. If radar is operating while the aircraft is on the ground, it can be hazardous for varying distances (depending on aircraft type). Some radars are dangerous up to 75 feet in front of the airplane and in a radius of up to 90 degrees on either side of the aircraft's nose. Pilots are trained to shut off radar when on the ground. In the event of a crash, it is unlikely that radar will be functional. The safest practice, however, is to avoid standing in front of the aircraft, not only because of radar, but also, as previously discussed, because of dangers from weapons systems and jet engine intake or propellers.

chaff
deters missiles launched at aircraft from the ground or from other aircraft

liquid oxygen (LOX)
oxygen that has undergone a cryogenic process that freezes the gas, thereby compressing it so that it can be stored in a smaller tank or reservoir; extremely cold and hazardous

Protruding Devices

Working close to any part of an airplane poses the chance of being poked or cut by protruding objects, such as probes, antennas, gun barrels, and static electricity dissipaters, and pitot tubes. Study **Figure 3-23,** which shows the front of an airplane. Observe a protruding, tube-like object; this is a **pitot tube**. These tubes may be straight or bent (such as the one in the figure). Angled pitot tubes are located on the sides of the forward portion of an airplane as well as on the leading edges of wings. Pitot tubes are used to probe the atmosphere and provide readings to the aircrew. These tubes contain a heating system to keep them from

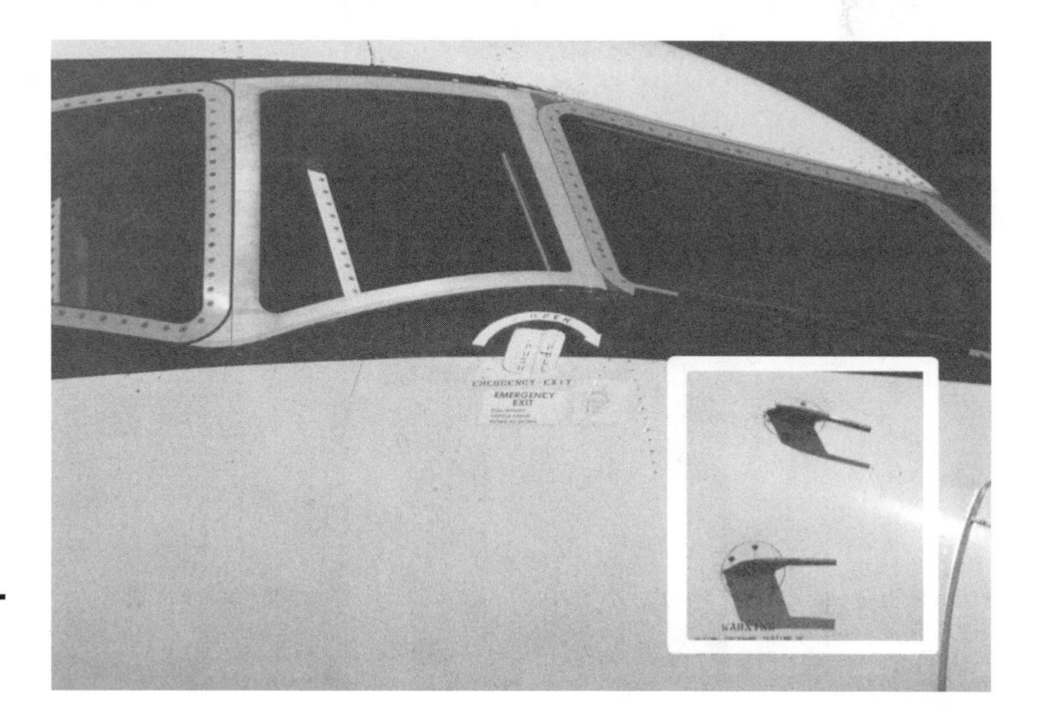

Figure 3-23 *Pitot tubes on a Boeing 757.*

pitot tube
a hollow, protruding tube that resembles a gun barrel and is attached to a wing or the nose of an aircraft; conveys the air speed information to display instruments on the control panel

freezing inside. Touching one with a bare hand shortly after an aircraft lands may result in static electric shock or burns. Also, rescuers may sustain an eye injury if they bump into this device.

Wing Surfaces

Figure 3-24 shows the dangers of walking on a wing surface. This can pose risk of injury due to falls. *Remember:* A wing is usually curved. The leading edge is rounded, like the front of a teardrop; the trailing edge tapers to a very thin edge. If you get too close to the edge, you are prone to falling off the wing. This is especially likely if the wing surface is wet from water, foam, fuel, hydraulic fluid, ice, or snow, or if your visibility is hampered by darkness or smoke. Avoid standing on control surfaces, such as flaps or ailerons, as they can bend or even break.

Warning
Entering the cockpit of a crashed aircraft is dangerous and must be accomplished deliberately. Never move any levers or switches unless you have been trained to do so.

Specialized Military Aircraft

When facing an incident with a military fighter aircraft, there are special hazards to be aware of when approaching and working around the scene.

Ejection Seats and Canopies Entering the cockpit of a crashed aircraft is dangerous and must be accomplished deliberately. Never move any levers or switches unless you have been trained to do so. In addition to the hazards posed by

Figure 3-24 *Gaining access to the aircraft over the wing. Note the shape of the wing surface; this hazard is further complicated by darkness.*

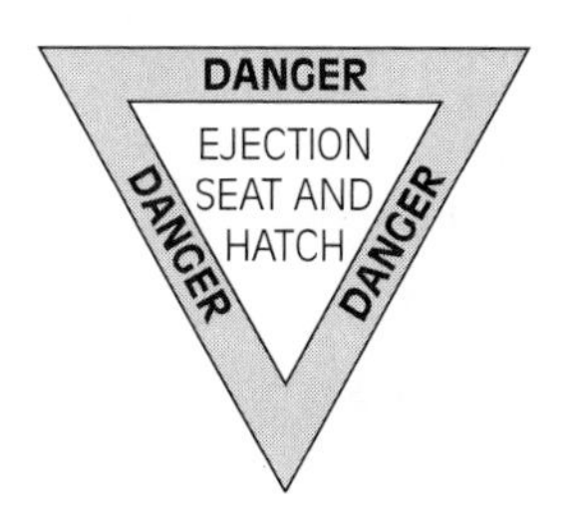

Figure 3-25 *An ejection seat warning. (Courtesy of the U.S. Air Force.)*

unidentified levers, some military aircraft are equipped with ejection seats and accompanying canopies or hatches, which may be triggered by the impact forces or by attempts to cut crewmembers lose. Crewmembers may be conscious and capable of safetying the ejection seat, or they may be able to instruct you how to accomplish this before you remove them from the cockpit. If there is no danger of fire or explosion, you may begin medical treatment, such as spinal immobilization, on crewmembers before they are removed from the cockpit.

■ Note
No external canopy or hatch jettison systems are engineered to activate ejection seats.

No external canopy or hatch jettison systems are engineered to activate ejection seats.

Figure 3-25 illustrates an important warning symbol that informs rescuers that the airplane is equipped with an ejection seat. **Figure 3-26** indicates where rescue controls are located to open a canopy, or if the situation warrants quick

Figure 3-26 *Emergency canopy access controls. (Courtesy of the U.S. Air Force.)*

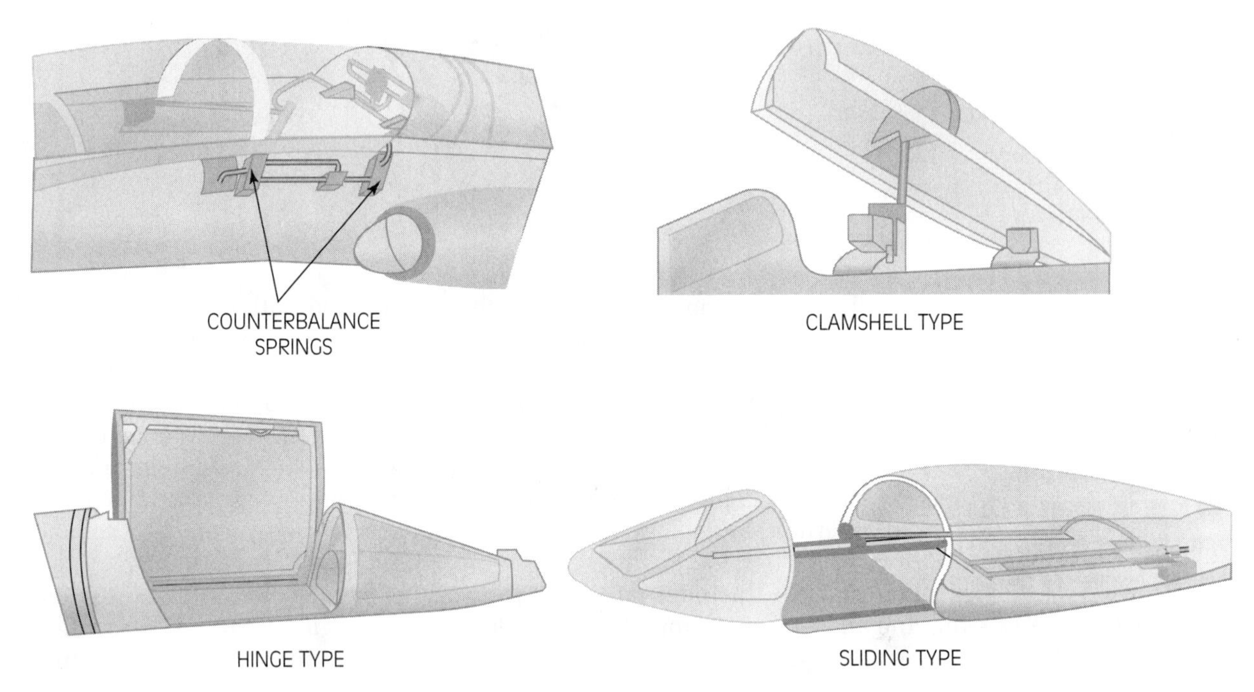

Figure 3-27 *Basic types of aircraft canopies. (Courtesy of the U.S. Air Force.)*

access, jettison the canopy, escape hatches, or windows; or how to open other means of quick access or escape routes for aircrew and/or passengers. Military aircraft are likely to bear dotted outlines of a designated "cut-in area."

Figure 3-27 shows the three basic types of canopies:

1. *Clamshell:* hinged, opens like a clamshell; the doors move up and back
2. *Hinged*: opens sideways
3. *Sliding:* slides on tracks toward the rear of an aircraft

Cutting into a canopy is a particularly hazardous operation. If you know for sure that trained airport crews are en route, and there is no danger, you may elect to initiate protective actions while awaiting their arrival. No two incidents are the same. When you are in the field, the circumstances may require you to make difficult judgments that only you can make, based on the information at the scene.

If you do decide to cut into a canopy, follow the instructions located close to the "rescue arrow" symbol on the aircraft. Instruct people to move far away. Pull the jettison lanyard, or handle, and run away as quickly as possible. When

Warning
When jettisoned, most canopies travel up and toward the rear (usually the right rear) of an aircraft.

jettisoned, most canopies travel up and toward the rear (usually the right rear) of an aircraft. Once the canopy has landed, resume your rescue operation.

You may determine that the canopy should be cut for several reasons, the most common being failure of the canopy to jettison. Cut close to the edges of the canopy frame (see **Figures 3-28** and **3-29**). A charged fire hose must be in proximity, although this is dangerous, as the footing may be uncertain. Other factors contributing to danger are weather, bad lighting, or surfaces that are slippery due to foam, hydraulic fluids, or spilled fuel.

Move people clear of the danger zones, which include the bottom rear of fighter and attack aircraft, before you begin cutting into a canopy. This is to prevent injury, not only from the potentially firing ejection seat and canopy, but also from flare and chaff dispensers, and from barrier-arresting hooks, which can drop quickly and violently from their storage at the bottom of the rear of the aircraft.

Danger
Rescue personnel may sustain burns from the rocket motors or from spilled fuel that has ignited. Hatches can be jettisoned externally, just like canopies that have this option.

Figure 3-30 shows that danger is also posed from control surfaces. Once the engines on some aircraft are shut down, these control surfaces may drop suddenly and with violent force. This can result in serious injury or death.

All ejection seats in U.S. military aircraft are capable of launching the air-crewmember high enough aloft for the parachute to successfully deploy (the exception is the downward-firing seats on the B-52). If an ejection seat is triggered, hazards are posed by the seat and the air-crewmember as they are propelled up.

AIRCRAFT ENTRY
4. CUT-IN

WARNING

Do not use portable gas rescue saw in an explosive atmosphere. This may cause an explosion and/or fire resulting in injury or death to pilot and rescue personnel.

a. Using portable gas rescue saw, cut out left or right side panel by cutting along inside edge of canopy frame on all four sides of panel.

b. Lift out panel.

NOTE:
Use a 12 inch diameter metal blade with carbide tips 3 and 1/8 inch pitch.

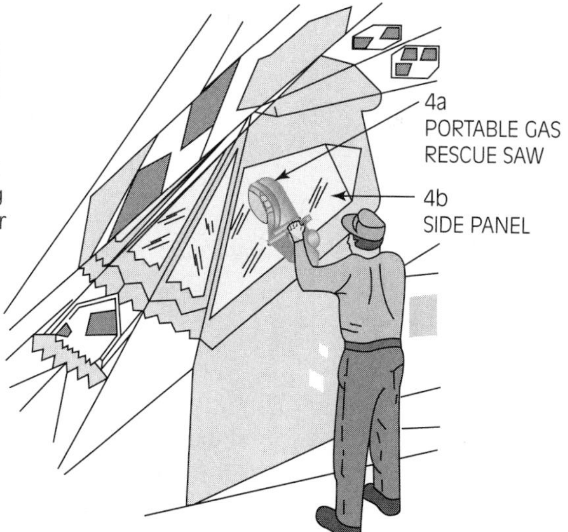

Figure 3-28 *Cutting an angulated aircraft canopy. (Courtesy of the U.S. Air Force.)*

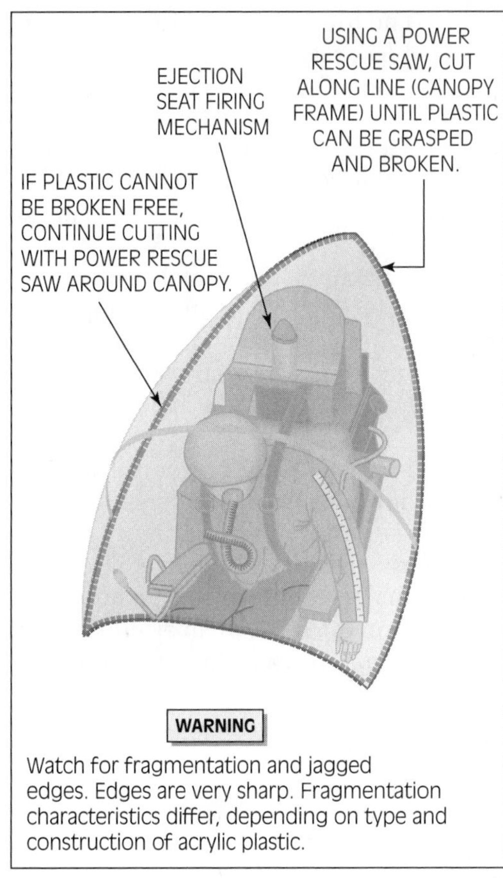

FORCIBLE ENTRY INTO PLASTIC CANOPIES

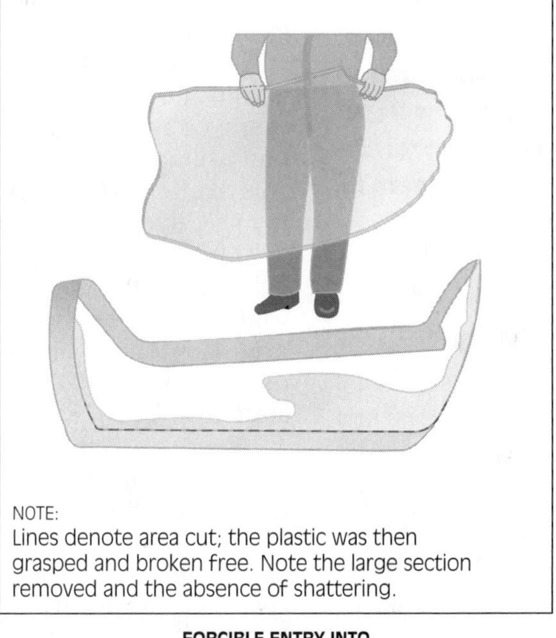

FORCIBLE ENTRY INTO A PLASTIC CANOPY

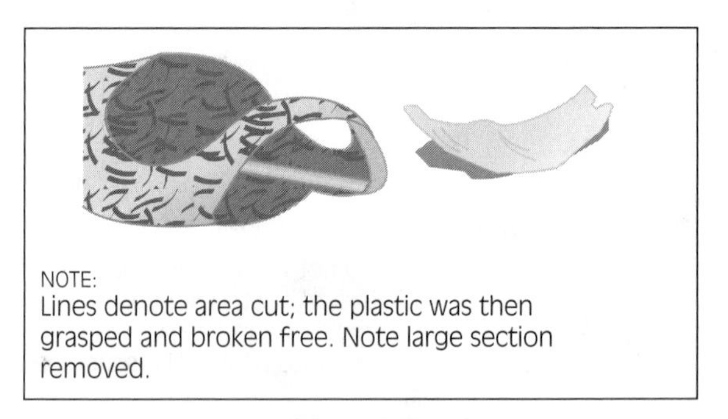

FORCIBLE ENTRY INTO A PLASTIC CANOPY EXPOSED TO FIRE CONDITIONS

Figure 3-29 *Cutting canopies. (Courtesy of the U.S. Air Force.)*

The inner barrel of the seat's catapult rocket remains with the aircraft, while the outer barrel stays with the seat as it is propelled out of the airplane. Rescue personnel may sustain burns from the rocket motors or from spilled fuel that has ignited. Hatches can be jettisoned externally, just like canopies that have this option.

MOVABLE SURFACES DANGER AREAS

WARNING

Personnel should stay clear of flight control surfaces when possible with the engines or APU running or external power and hydraulics applied. Danger areas are highlighted with the rudders posing the least hazard. Failure to disregard danger areas can result in injury or death.

WARNING

The arresting gear is located far centerline aft of the shoe hook. It is pneumatically extended and retracted. Injury or death to personnel can occur during operation.

WARNING

The safing pin prevents the cable movement required to actuate the arresting gear to extend. Personnel should stay clear of arresting gear at all times. Injury or death to personnel can occur if the hook safing mechanism fails.

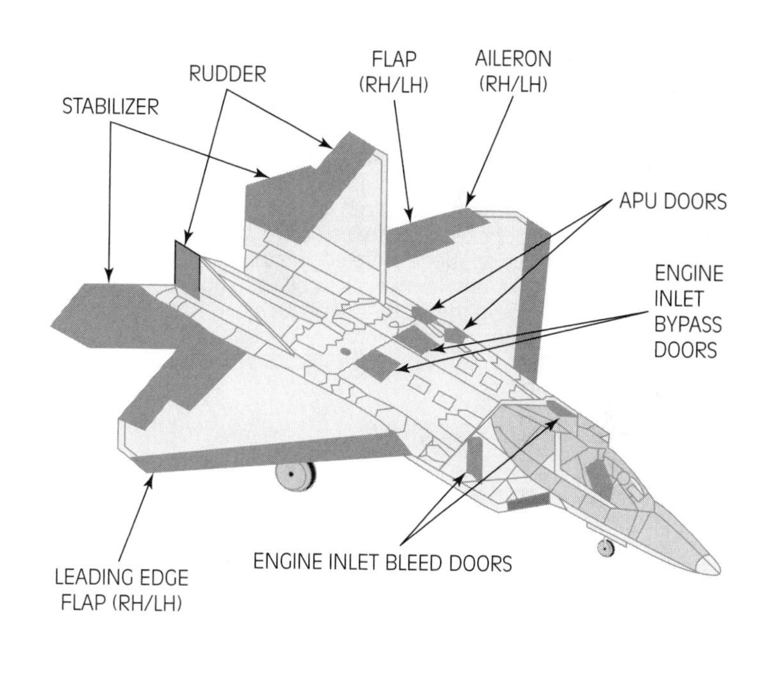

Figure 3-30 *Control surface danger zones around a fighter aircraft—keep people clear. (Courtesy of the U.S. Air Force.)*

Figures 3-31 through 3-34 are for *academic and informational purposes only*. Military ARFF crews spend long periods during initial and recurring training sessions practicing the correct techniques to safety ejection seats. The seats most commonly used in American military aircraft are ACES and Martin Baker. Each manufacturer makes variants of these seats based on the aircraft, such as, newer production variants of Martin Baker and ACES seats. Other nations may use different ejection seat variants.

! Warning **If crewmembers are unconscious, it may be easier to cut harnesses rather than try to remember how each restraining system functions.**

As mentioned earlier in this chapter, if crewmembers are awake and ambulatory, they probably can unfasten their restraining harness belts and exit the airplane on their own. Also, they may place the safety lever in the correct position so they can be extricated from the aircraft. If crewmembers are unconscious, it may be easier to cut harnesses rather than try to remember how each restraining system functions.

Figure 3-31 shows a variants of the ACES II ejection seat.

Figures 3-32 and 3-33 show a variant of a Martin Baker ejection seat.

Figure 3-34 illustrates process of removing the pilot from an aircraft equipped with a Martin Baker ejection seat.

SAFETYING EJECTION SEAT

WARNING

The seat is armed regardless of canopy position. Jettisoning the aircraft canopy prepares the ACES II ejection seat for ejection. Seat(s) can eject whether canopy is opened or closed. On two seat aircraft, both seats must be safetied before either can be considered safe. Extreme caution must be used not to inadvertently move the Ground Safety Lever from the SAFE position during aircrew extraction. DO NOT USE PITOTS FOR HANDHOLD DURING ANY TIME OF THE OPERATION.

1. NORMAL SAFETYING EJECTION SEAT

NOTE:
The Ground Safety Lever Safety Pin can be installed regardless of seat position.

a. Rotate Ground Safety Lever, located on left side of seat, UP and FORWARD, and install safety pin in pin receptacle at base of lever near pivot point. Pin faces forward. If safety pin cannot be installed, tape or tie Ground Safety Lever in UP position to prevent arming during extraction.

b. Install Safety Pin in the Emergency Manual Chute Handle. If Ground Safety Pin and Emergency Manual Chute Handle Pin are connected by one safety streamer, route Emergency Manual Chute Handle safety pin under aircrew's legs, otherwise extraction will cause entanglement with streamer.

2. EMERGENCY SAFETYING EJECTION SEAT

a. Rotate Ground Safety Lever, located on left side of seat, UP and FORWARD.

b. Insure Ground Safety Lever does not rotate downward and arm seat during extraction or movement of aircrew.

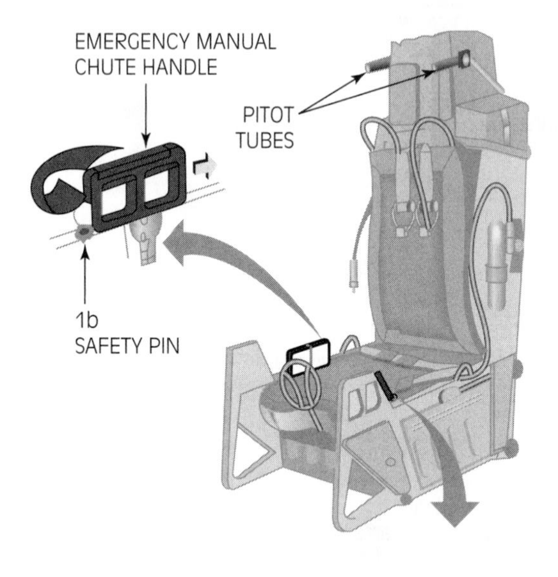

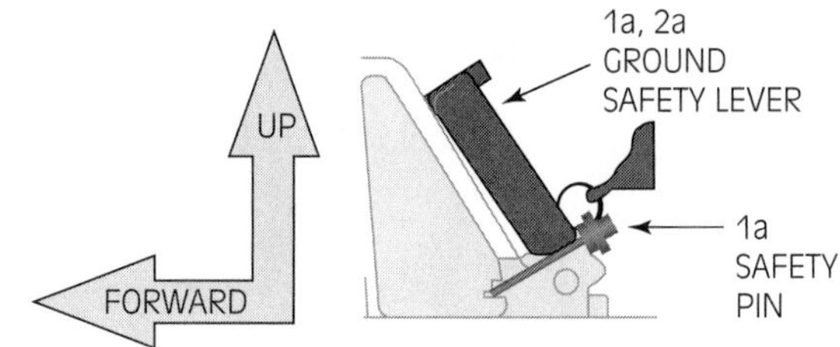

GROUND SAFETY LEVER SAFE POSITION

Figure 3-31 *Aces II ejection seat. (Courtesy of the U.S. Air Force.)*

PERSONAL PROTECTIVE EQUIPMENT FOR AIRCRAFT INCIDENTS

Personal protective equipment for ARFF differs from the equipment for structural, wildland, and shipboard firefighting, which is essentially the same. ARFF PPE is designed to combat blistering heat and the force of fuel-vapor explosions. Conversely, wildland firefighting clothing is much lighter and more flexible than

EJECTION SEAT SAFETYING MARTIN-BAKER SJU-5/A,

1. EJECTION SEAT SAFETYING-NORMAL AND EMERGENCY-SJU-5/A, 6/A MODEL

NOTE:

Immediately upon gaining access to the aircraft cockpit, if time permits and no hazardous conditions exist, proceed with seat safetying procedures.

WARNING

If ejection control handle is not fully seated, safety pin cannot be installed and safe/armed handle cannot be rotated to the fully locked position. An unsafe seat exists if the entire word "SAFE" is not visible on the safe/armed handle. If ejection seat is not in a safe condition, initiation may occur if ejection control handle is pulled. Proper procedures for resetting handle must be followed.

a. Insert safety pin into ejection control handle if handle is in first detent (stowed) position. If ejection control handle is not in stowed position, return handle to first detent (stowed position) by pressing handle into its housing and inserting safety pin.
b. Press button on top of manual override handle and rotate handle UP and AFT. The safe/armed handle will simultaneously rotate up and the entire word "SAFE" should be visible.

WARNING

In multi-seat aircraft, all ejection seats must be safetied.

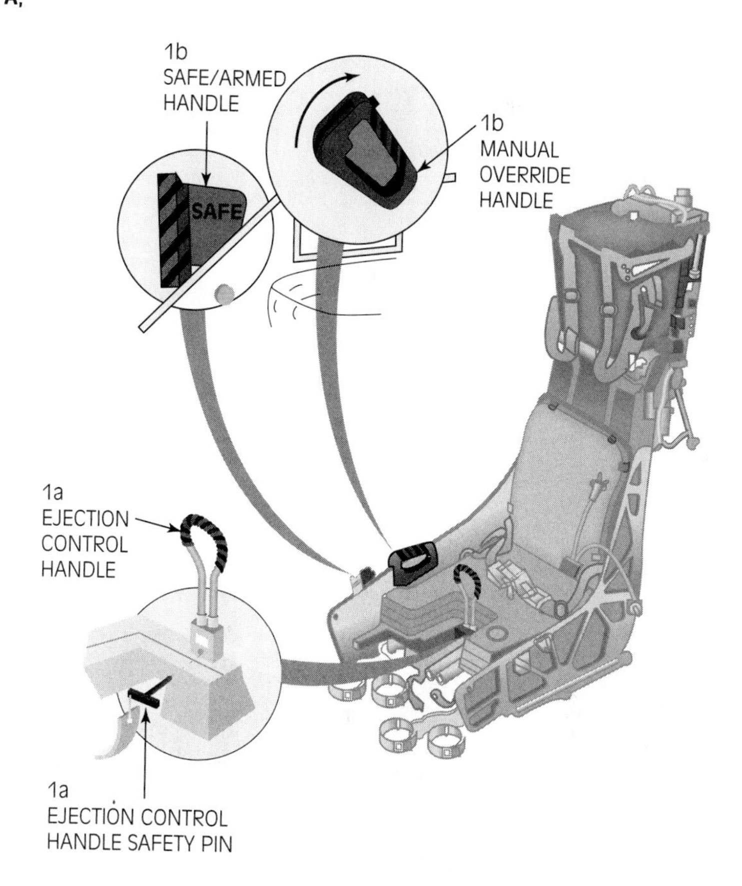

Figure 3-32 *A Martin Baker ejection seat. (Courtesy of the U.S. Air Force.)*

ARFF firefighting clothing. The heat-reflective abilities of ARFF-approved ensembles are sacrificed in wildland firefighting equipment, so that the clothing is light enough to permit firefighters to hike through rough terrain to access and fight a wildland fire.

personal protective equipment
intended to protect rescue personnel from injuries; includes head gear, special boots or shoes, clothing, and respiratory masks

USING YOUR PERSONAL PROTECTIVE EQUIPMENT

Normally, the preferred **personal protective equipment** for aircraft incidents is an aluminized-coated protective gear. This gear is similar to structural firefighting ensembles in that the aluminized coating reflects radiant heat. When this

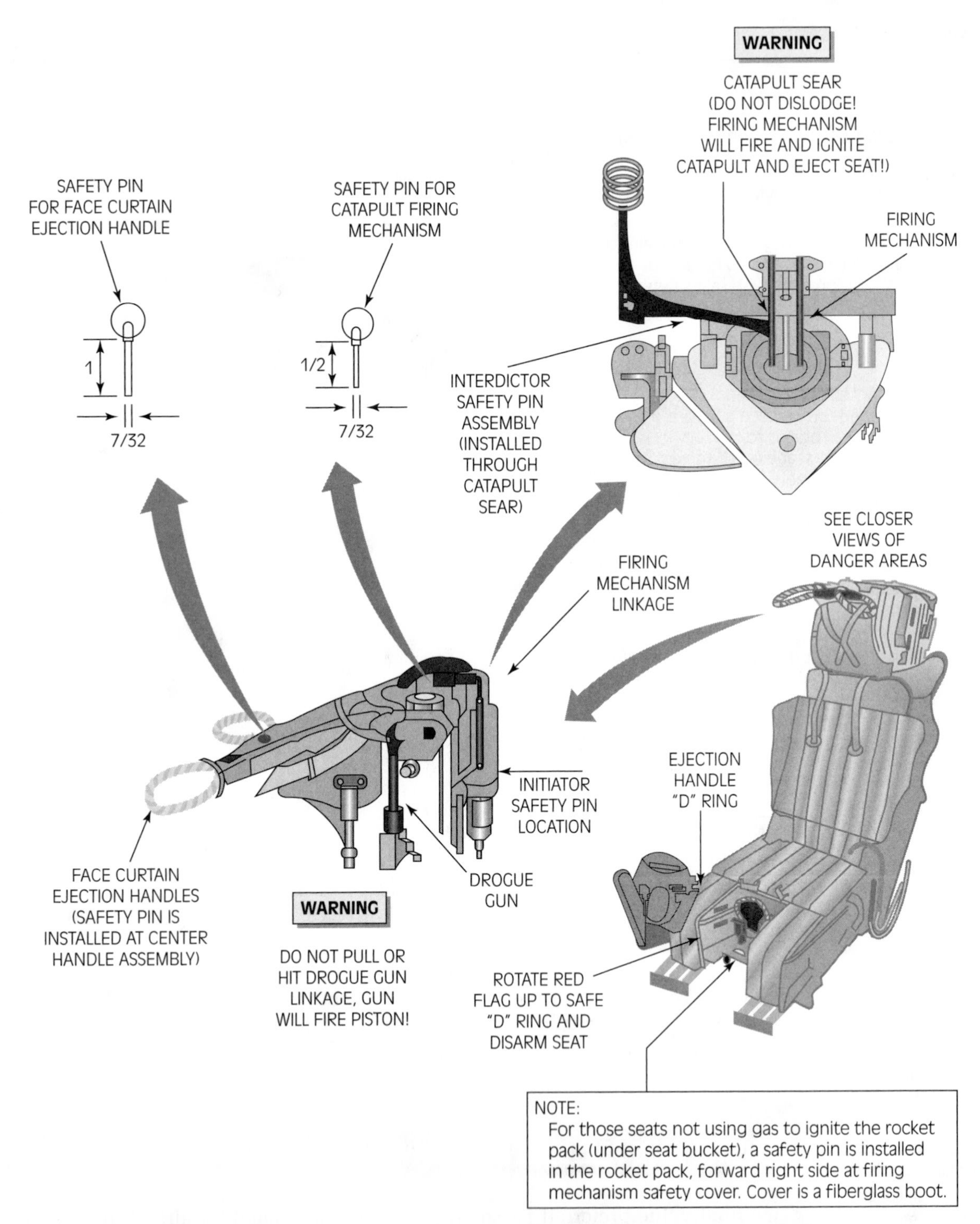

Martin-Baker Ejection Seat Showing Disarming Safety Pins and Red Flag

Figure 3-33 *Another look at a Martin Baker ejection seat. (Courtesy of the U.S. Air Force.)*

AIRCREW EXTRACTION

1. AIRCREW EXTRACTION

NOTE:
The crewmember is attached to the seat by the use of an integrated harness and leg restraints. Additionally, the oxygen/communication lead is attached to the survival kit. If the crewmember is wearing an anti-G suit, a hose will be attached to an outlet on the LH console.

a. To remove oxygen mask: Pull down release tabs on either side of crewmember helmet mask.
b. To disconnect the oxygen/communication lead at the survival kit on the left aft side of seat: Grasp knurled fitting on hose and pull up to disconnect.
c. To disconnect the anti-G suit: Pull anti-G suit hose from left seat connection.
d. To disconnect leg restraints: Release leg garters by applying pressure to tabs on both sides of each quick disconnect.
e. To disconnect restraints: Release two lap belt, then two shoulder harness koch fittings.

2. EMERGENCY RELEASE

a. Press thumb button on forward part of manual override handle, located on right side of seat, and rotate handle aft. This positions the safe/armed handle UP in safe position and releases lower leg restraint lines. However, the parachute and survival kit will remain attached to crewmember.

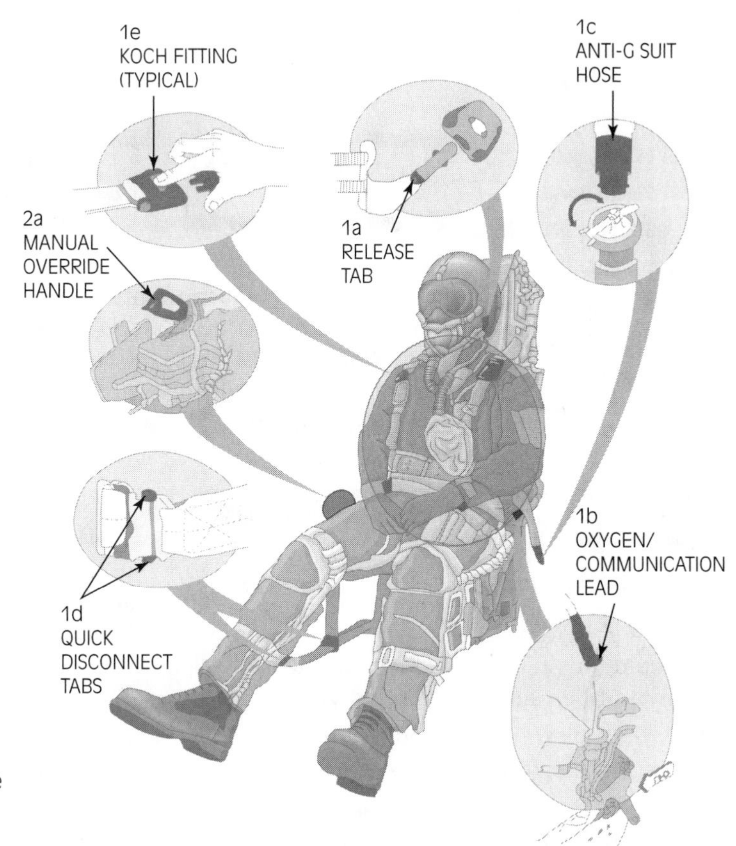

Figure 3-34 *Leg restraints, harnesses, hoses, and buckles must be unfastened before a pilot can exit his/her attack or fighter aircraft. Most high-performance aircraft equipped with ejection seats, such as the one pictured, have leg-restraint straps to minimize aircrew injury in the event of ejection. (Courtesy of the U.S. Air Force.)*

ensemble is brand new, it may reflect 90 percent of heat, but as the surface fades with age and exposure to the elements, its heat-reflective ability decreases. Unless you are actively involved in an airport environment, a petroleum fire brigade, or another specialized firefighting activity, you may not have this equipment in your inventory. The structural firefighting ensemble provides only limited protection from the high-heat environment of aircraft fires. Aircraft fuel fires may be extremely hot. If the inside of your firefighting clothing becomes wet, you may suffer steam burns. At least, it is best to wear pants, T-shirt, and cotton socks beneath firefighting clothing to absorb moisture.

In addition to any protective gear you have, you must use a self-contained breathing apparatus within the accident cordon, even after the fire has been

• Caution
Aircraft fuel fires may be extremely hot. If the inside of your firefighting clothing becomes wet, you may suffer steam burns. At least, it is best to wear pants, T-shirt, and cotton socks beneath firefighting clothing to absorb moisture.

extinguished, because toxic fumes will off gas until they can be mitigated by the site restoration team. Once the cordon is established, set up air-monitoring equipment to ensure all responders are kept safe from toxic gases generated by materials being consumed or involved in fire.

NFPA REGULATIONS FOR PERSONAL PROTECTIVE EQUIPMENT

The National Fire Protection Association (NFPA) has developed national consensus standards for firefighting ensembles used in aircraft emergencies. Review source document NFPA 1976 to understand compliance with these regulations. NFPA Standard 1971 was created to establish guidelines for protective ensemble for structural firefighting. Also refer to this standard for guidance and compliance requirements.

SUMMARY

This chapter explored common hazards found at the scenes of aircraft accidents.

- You should be more aware of the common and less common dangers posed by an aircraft in an accident situation. These include, but are not limited to, dangers from the electrical systems, hydraulics, and pressurized vessels.
- Military aircraft pose additional dangers, because they often are equipped with weapons and munitions, ejection seats, and jettison systems.
- Additional hazards are presented by the combustion products generated by standard and advanced aircraft construction materials. These include, but are not limited to, toxic vapors found in smoke generated by burning cargo, baggage, and other airplane contents.
- Dangers from declared and undeclared hazardous cargo, extraordinary systems, or specialized aircraft may also be encountered during a rescue.
- The use of personal protective equipment was briefly discussed, as were the NFPA regulations governing PPE use.
- Finally, remember that responders' vigilance of a potentially volatile situation can help ensure that rescue operations are conducted safely for rescuers, bystanders, and victims of the incident.

KEY TERMS

Aqueous film forming foam (AFFF) Fire-extinguishing agent that contains fluorocarbon surfactants and spreads a protective blanket of foam that extinguishes liquid hydrocarbon fuel fires by forming a self-sealing barrier between the fire and fire-sustaining oxygen. Note: Application of foam % means the percentage of concentrate in the final solution (e.g., 6% means 6% foam and 94% water).

AVGAS High-octane gasoline with a flash point of –36° F.

Boiling liquid expanding vapor explosion (BLEVE) Often called a "blevie," this dangerous event results when a closed container of liquid is exposed to excessive heat or flame impingement. The result is catastrophic container failure that often results in fragments of the container dispersing with violent force.

Buddy store This aerial refueling system is used by the U.S. Navy S-3 Viking airplane. It uses a modified, externally mounted fuel tank with the drogue-parachute type of refueling system. It is attached to a weapons/bomb rack on aircraft and has been successfully tested by Boeing Aircraft, which used an A/F-18 Super Hornet jet to refuel other aircraft in flight.

Chaff This material deters missiles launched at aircraft from the ground or from other aircraft. During the Second World War, it was developed as a countermeasure, a method of confusing radar during combat. Aircraft (or other targets) disperse pieces of aluminum foil, metallized glass fiber, or plastic to form a cloud. Rescue personnel should preplan various aircraft to learn where chaff dispensers are located on them.

Cockpit shutdown procedures This term is often used to mean "emergency shutdown procedures," however, it is the methodical shutting off of aircraft engines, fuel pump switches, electrical power switches, hydraulic systems, and so on.

Emergency shutdown procedures These are methodical and often sequential steps for shutting down an airplane. They include shutting the aircraft throttle (or throttles) to the idle, then "off" position; turning off electrical systems; safetying ejection seats (if possible); turning off fuel selection switches, and so on.

HE (pronounced "H-E") An abbreviated term for high explosives, which are contained in ammunition, bombs, and missiles or cannon shells.

International Air Transport Association (IATA) This international industry trade group of airlines is based, with **ICAO,** in Montreal, Quebec, Canada. Today, IATA has more than 270 members from more than 140 nations. This organization helps airline companies in such areas as pricing uniformity and establishing regulations for the shipping of dangerous goods, and it publishes the all-important *IATA Dangerous Goods Regulations* manual. This manual is recognized worldwide as the field source reference for airlines shipping hazardous materials referred to in aviation as dangerous goods.

International Civil Aviation Organization (ICAO) A specialized agency of the United Nations that is responsible for developing international rules governing all areas of civil aviation. Rules relating to transportation of commodities must comply with Title 49 CFR. The ICAO's safety responsibilities include a regulatory framework, enforcement and inspection procedures, and, when necessary, corrective measures associated with airworthiness of aircraft, airport safety, personnel licensing, and international aviation rules.

Jet A fuel Kerosene-type fuel with a flash point of 110° F to 115° F.

Jet B fuel Kerosene-type fuel with a flash point of –16° F to 30° F.

Jet engine intake The portion of a jet engine where air enters in at great velocity. The engine intake poses a potential hazard for responders.

Life safety This term refers to the protection of human life.

Liquid oxygen (LOX) Oxygen that has undergone a cryogenic process that freezes the gas, thereby compressing it so that it can be stored in a smaller tank or reservoir. LOX is extremely cold and hazardous (see Section 3).

Magneto This device is found in aircraft piston-type aircraft engines and generates electric current from a magnet that spins as the motor crankshaft operates. If the aircraft's electrical system fails, the magnetos ensure that the engine spark plugs get electricity. Engines usually contain two sets of magnetos.

National Institute of Safety and Health (NIOSH) Part of the U.S. Department of Health and Human Services, NIOSH operates within the Centers for Disease Control and Prevention (CDC) and is the federal agency responsible for conducting research and making recommendations for the prevention of work-related injury and illness.

The *Occupational Safety and Health Act (OSHA)*, established in 1970, created NIOSH and OSHA. Different states, however, take different approaches to legislation, regulation, and enforcement of these regulations.

Similar worker-safety provisions also are enforced in Canada and in many European Union countries, which have enforcing authorities to ensure that the basic legal requirements relating to occupational safety and health are met ensuring good OSH performance.

In Canada, the Canadian Centre for Occupational Health and Safety (CCOHS) was created based on the belief that all Canadians had "... a fundamental right to a healthy and safe working environment." CCOHS is mandated to promote safe and healthy workplaces to help prevent work-related injuries and illnesses.

The European Agency for Safety and Health at Work (EASHW) was founded in 1996. In the UK, health and safety legislation is drawn up and enforced by the Health and Safety Executive under the Health and Safety at Work Act of 1974.

Ordnance This term refers to ammunition, bombs, rockets, or other explosive materials that may be carried on military aircraft.

Personal Protective Equipment Equipment intended to protect rescue personnel from injuries. It includes head gear, special boots or shoes, clothing, and respiratory masks.

Pitot tube A hollow, protruding tube that resembles a gun barrel and is attached to a wing or the nose of an aircraft. It may be shaped like the letter "L." It measures RAM air pressure and translates those measurements into air speed. (This is discussed in detail later.) The tube, which is equipped with an internal heating device to prevent freezing, conveys the air speed information to display instruments on the control panel.

Thermal runaway condition This is an electrochemical reaction that causes a battery to overheat, release toxic vapors, spew electrolyte, and very likely explode. An aircraft suffering this problem must land as soon as possible.

REVIEW QUESTIONS

1. Jet fuel is similar to what other type of fuel?
2. What are factors that contribute to a rescuer falling off a wing surface?
3. What are the hazards posed by Skydrol LD-4 and 5?
4. When a volume of liquid oxygen changes to a vapor, how much does it expand (i.e., what is the expansion ratio)?
5. What is our biggest safety concern when liquid oxygen has spilled onto an asphalt surface?
6. An aircraft engine magneto is very much like an early automobile or a lawn mower engine magneto. What is it designed to create?
7. Study **Figure 3-35.** These cockpit diagrams are from various commercial and general aviation aircraft and have several things in common:
   - What do they "arm"?
   - What do they usually shut off?
   - Can fire extinguisher switches discharge an onboard fire extinguishing agent if the electrical power has been totally shut off?

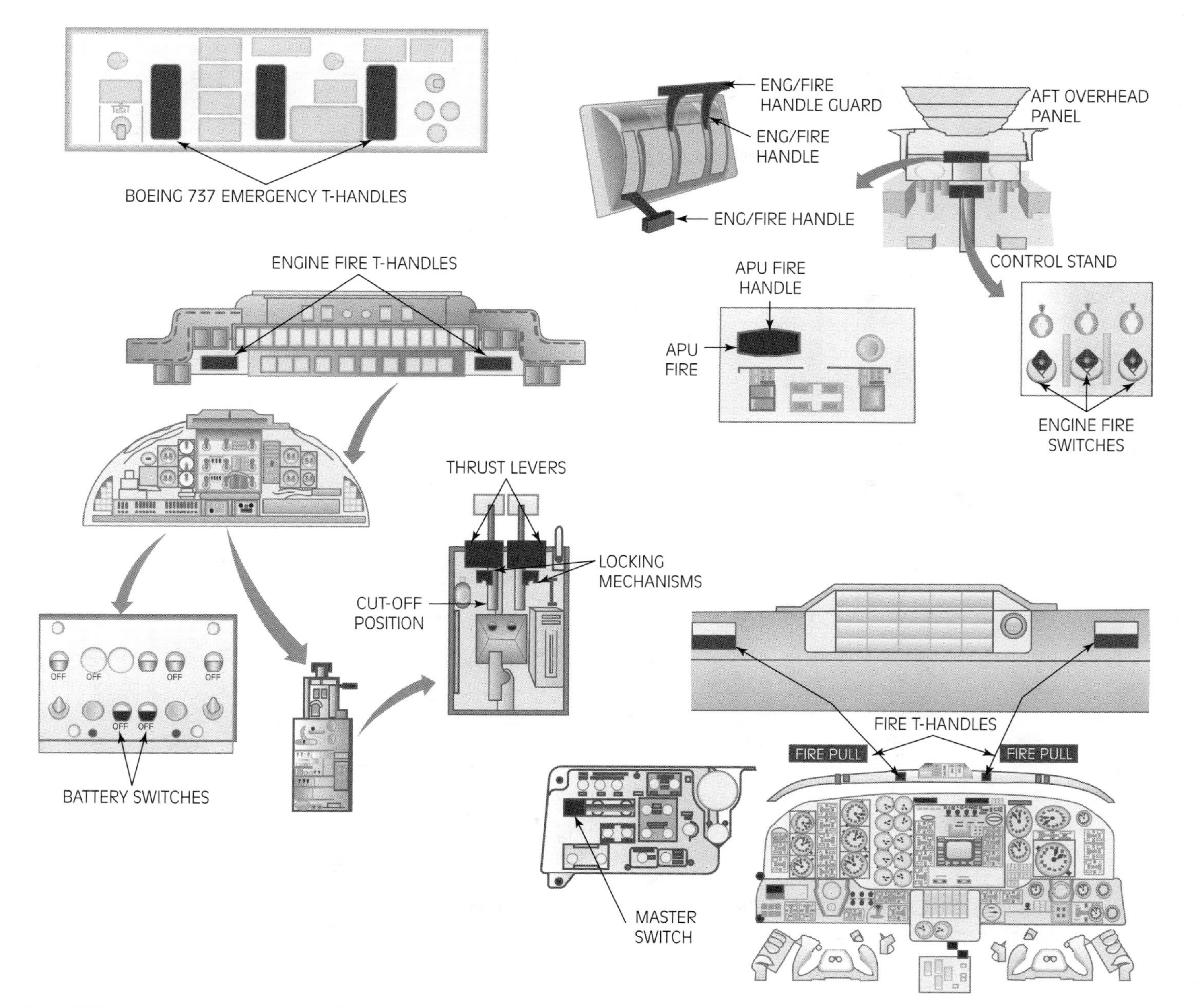

Figure 3-35 *A selection of commercial and general aviation aircraft control panels and "T" handles. (Courtesy of the U.S. Air Force.)*

8. What direction do most canopies travel when jettisoned?
9. What is the health hazard of staring directly at burning magnesium?
10. T F A person standing 15 feet from the front and 90 degrees to the side of an operating jet engine can be ingested (sucked) into the engine's intake.
11. Always assume HAZMAT is on board an airplane as part of the cargo or baggage. If the aircrew of a transport aircraft is incapacitated, or if the aircraft has undergone a catastrophic accident, what should rescuers look for that contains information about dangerous goods?
12. List the three basic types of canopies.
13. What is the three-letter acronym (nickname) for liquid oxygen?
14. What is the two-letter abbreviation for hazardous materials that are transported by aircraft and are defined as substances capable of posing a significant risk to health, safety, or property when transported by air?

STUDENT EXERCISES

1. International and federal law forbids certain materials from being transported by air as baggage or cargo. Search the NTSB Web site, newspaper articles, or other resources and cite one case history for your study group. Provide the following:
   - name of forbidden material carried
   - suggested placard
   - Material Safety Data Sheet (MSDS) information
2. The following example is based on an actual incident: You are assigned to a Hazardous Materials Crew at an off-airport fire station. You are dispatched to the airfreight terminal because a wooden crate was unloaded and is leaking. A trail of liquid leads to the cargo hold of a passenger jet 300 feet away. After conferring with the aircraft pilot and the airline representatives at the cargo terminal, you learn the following:
   - The aircraft flight originated at a small airport in Alaska.
   - The wooden crate contains the human remains of a drowned fisherman.
   - This fluid has contaminated the cargo floor of the jet, as well as some baggage.

   Describe your *concerns*, *considerations*, and *actions* based on your job assignment (HAZMAT, public health, airline, fire officer, FAA, and so on).
3. At an aircraft crash site, you notice several small drums, bearing the double "This End Up" labels, scattered about. Most of these drums are leaking an unknown product that is mixing with other unknown cargo and baggage. National Transportation and Safety Board personnel have not arrived. If you were the incident commander, what actions would you take in the interest of safety, while not compromising accident investigation evidence?
4. You are called on to cut the canopy on a fighter type jet aircraft. Describe how you would safely accomplish this.
5. Discuss the hazards posed by hydraulic lines and pressure vessels in a post-crash or fire environment.

RESPONSE CONSIDERATIONS

Learning Objectives

Upon completion of this chapter, you should be able to:

- Know your response area.
- Understand how to establish incident command.
- Understand unified command.
- Understand how to answer the question: What's your plan?

INTRODUCTION

No one can predict with certainty where or when an emergency will happen. As an emergency responder, you are the answer to someone's problem. The foundation of emergency response is to take control of someone's unfortunate or tragic situation and provide the most favorable outcome for the event. To achieve this goal, you must plan your response considerations for myriad emergency situations, and then regularly reevaluate those plans. Preplanning for and understanding your response options and the tools you need for incident command are essential steps toward effectively controlling any incident. Begin by acquiring detailed familiarity with and an in-depth understanding of your local response area.

KNOW YOUR RESPONSE AREA

Whether it happens on a country road, an airport runway, or in a neighborhood, an aircraft accident or fire requires rapid response. Before you can respond to an accident, however, you must be able to find it.

Commercially available GPS mapping systems can be installed in most vehicles and are quite helpful when you need to locate the site of a downed airplane. Unfortunately, this equipment often is unreliable and breaks down. Large-scale natural disasters, for instance, often disable satellite communications systems, which can include GPS mapping systems.

In such a situation, you must be able to fall back on your knowledge of the area. Obtaining this knowledge requires you to traveling the area in depth, so that you become familiar with every road, bridge, and waterway it contains, as well as having a solid sense of the region's general terrain. Always have hard-copy maps of the region on hand, including community grid maps. In addition, work with your county, city, airport, state, and local military base officials to ensure that all agencies are working from the same maps.

Roads

Knowing your response area well prepares you to use alternate roads, such as smaller, secondary roads, and to know which of them can accommodate your apparatus. Being familiar with the roads in your response area enables you to decide quickly on the best response routes, primary and secondary, to the accident site. Several factors must be considered when determining the appropriateness of various routes, including weight restrictions for heavy apparatus and "pinch points" (narrow places) that hinder the movement of responders into and out of an area. Vehicle width may require considerations of single-lane traffic operations, which must be coordinated with traffic control authorities or local law enforcement. Landing zones for medevac helicopters, bridges, and parking/staging areas must be carefully coordinated with law enforcement personnel, aircrews, and emergency responders.

Bridges also are important considerations in your preplanning. Which of the bridges in your region that you might need to traverse can support the weight of your vehicle? Are any bridges (or roads) too narrow for large fire and rescue apparatus? Are the bridges' truss works high enough to allow your larger vehicles to pass under them? Also, know the heights of highway overpasses, so that you know which of them will accommodate various emergency vehicles.

As you ride around your initial response area, make mental notes and take written notes of what factors must be considered in the event of a large-scale incident.

Waterways, Lakes, Rivers, and Reservoirs

Aircraft accidents occurring on land pose enough hazards to responders, but accidents occurring in water are even more hazardous. The first issue you face in a water-based rescue is approach: How do you get to the scene? Traveling to the site may be complicated by narrow en route bridges and roads. Once you have arrived on scene, the next issue is fire suppression: Fuel products float, thus creating a large area of involvement and complicating fire attack strategies. Rain water runoff also affects spilled-fuel firefighting tactics, especially if the water is flowing with a current. Hazardous materials, like fuel, may be washed downstream or spread widely across the area, endangering responders, victims, and bystanders, and probably causing pollution problems. Also, the containment and reduction of contamination from biohazards and other hazardous materials must be dealt with and may cause problems far beyond the impact area. Patient recovery may be hindered if responders cannot reach survivors without boats. Finally, any evidence useful to the investigation team could easily be compromised or lost in water.

Terrain

If an aircraft accident occurs in a rural setting, it may cause a wildland fire, which will additionally tax your resources. In this case, you may require the help of mutual aid units. Unfortunately, given the remote setting, their arrival may be delayed. When fighting wildland fires, the **incident commander** (IC) may need to request additional water-tender fire trucks to shuttle water to firefighting vehicles, because rural settings are unlikely to have sufficient, or even any, fire hydrants. When requesting such vehicles, it is critical—as always—to know the vehicles' specifications (weight, height, and width).

incident commander (IC) person who is designated as being responsible for all activities associated with an incident, including the development of tactical plans, strategies, and the utilization of resources

Hilly terrain makes it difficult for vehicles to reach the scene, especially if it contains steep grades, canyons, or gullies. Prevent vehicle rollovers on the sides of hills by knowing your vehicle's limitations regarding the degree of incline it can safely traverse. Accessing swamp areas requires off-road fire vehicles. In sandy or muddy terrain, some air can be let out of vehicle tires to provide better traction. Know which of your vehicles are designed with off-road capabilities and how effectively each vehicle type moves through mud and sand, on steep hills, and so on.

Be especially alert and cautious during your approach to a rural incident scene. Deceased victims and survivors may be either lying on the ground or walking about. For example, if smoke is engulfing a scene, driving fast through it may endanger victims and volunteer rescuers, or place your vehicle into a ditch or similar obstacle.

When dealing with a military aircraft accident, watch for and avoid externally mounted fuel tanks or munitions; these also may be scattered about the crash area. Look for signs of high-explosive material (HE); this may have partially burned, or melted and resolidified. This material may resemble a "mud pie," having taken on the color of the surrounding dirt, perhaps with some discoloration. Never drive over, step on, or otherwise disturb this material, as it is shock sensitive.

Danger

Look for signs of high-explosive material (HE); this may have partially burned, or melted and resolidified. This material may resemble a "mud pie," having taken on the color of the surrounding dirt, perhaps with some discoloration.

Part of becoming familiar with your surroundings is being aware of what areas might serve as emergency landing areas. As you explore the elements and terrain of your region, look for roads and fields (such as farmland) that a pilot experiencing flight problems may use as an emergency runway, such as an interstate highway, which may become an emergency landing strip at any time. Note the access roads to these potential landing sites: are they mud bogs during certain months of the year? If so, this will affect your abilities to access the sites at those times. In open areas, a pilot's efforts to make an emergency landing may be more or less successful depending on the area's size and terrain.

Facilities, Departments, and Co-Agency Capabilities

Once you find these potential emergency landing places, determine which facilities (buildings, neighborhoods, industrial parks, and so on) may be involved in such a landing. Find out what services are available to respond to an off-airfield accident or emergency landing. Does the local airport have a full-time aircraft rescue firefighting department? How close are other fire departments? Does the airport have trained security personnel capable of gaining and maintaining scene control? Does it have personnel and equipment capable of containing large amounts of hazardous materials? Does it have heavy-duty rescue equipment, as well as other heavy-duty equipment, such as road graders, dump trucks, cranes, and so on that may be needed for cutting emergency access roads for post-incident debris removal? These are just a few of the questions you should pose to local airport managers, private aircraft owners, and pilot associations. These topics need to be considered *before* an emergency happens. When you preplan with potential responding agencies, however, be sure to identify their capabilities as well as their limitations.

Know Your Resources

In the early morning hours, at 0300, it is difficult to determine what resources are available to help you successfully manage the outcome of the aircraft accident. Certainly, making a radio call or calling 911 will get you help, but other numbers, such as the one for the nearest airport or military base, may be difficult to obtain quickly in the midst of a crisis.

To avoid this situation, take the time to become aware of all the potential resources in your region and to determine which of these are available for response. Assistance for an aircraft accident is available from a wide array of agencies that can provide equipment and human resources. Private construction contractors and local governments may have ready access to special heavy equipment. Know which agency has jurisdiction in areas within and adjacent to your community. Examples include the U.S. or State Forest Services, Bureau of Land Management, Department of Food and Agriculture, Bureau of Indian Affairs, Department of Energy (for wildlands), state or national parks, or special-use properties. The Coast Guard may be an asset for accidents by navigable waterways.

Additional key resources include the Red Cross, Salvation Army, and other local community groups. Disaster-aid groups in your area also can provide services as needed.

Make a list of telephone numbers for every agency and facility you might need to call on during an emergency. Your communications center should have the telephone numbers of the federal, regional, state, county, and local emergency operations centers. Many military bases have a command post that is staffed twenty-four hours a day. Passenger airports also are likely to have emergency operations centers. Update this list frequently, to include new resources as they become available and to delete any resources that have become unavailable. In addition to daytime and nighttime telephone numbers, obtain the names of contact persons and their alternates. Include these peoples' personal cell phones, if they permit it. Most contact people will provide these numbers to emergency response agencies, if you agree to keep the numbers confidential information. Your response plans should include estimated response times for each resource for day and nighttime emergencies.

ESTABLISHING INCIDENT COMMAND

"When in command, *command*!"

Some agencies refer to the oversight of an accident scene as "incident management," others may call it something else. Whatever terms are used, however, all refer to the same thing: *managing* the operation. The one element that determines the success or failure of an emergency response is establishing **incident command**. Incident command is especially critical in an aircraft accident,

incident command especially critical during aircraft accident rescue, because crew and passengers are encased in these craft and surrounded by large volumes of flammable aviation fuel, as well as baggage and cargo that may contain hazardous materials

because crew and passengers are encased in these craft and surrounded by large volumes of flammable aviation fuel, as well as baggage and cargo that may contain hazardous materials. Under these daunting circumstances, prompt, effective actions to contain an accident scene and rescue as many victims as possible are possible only with preplanning and proper on-scene management.

Only one person should direct and manage the entire scene of an aircraft accident. You can announce incident command and identify who is in charge by using a radio announcement, a flag, a flashing light, or even wearing a vest. During most emergencies involving fire departments, people look for the person wearing the white helmet. When other responding agencies with a role in the operation arrive, the incident commander should establish a unified command system when possible. Trained responders know that they should have one, primary boss using an incident command or incident management structure, who, in concert with a unified command, delegates authority to his or her "helper (secondary) bosses," in compliance with National Incident Management System (NIMS). All emergency responders should be trained in and know NIMS regulations and procedures, as well as any other organized response plans that can be applied at the local level for smaller events, to enable this multi-agency system to work effectively.

Cooperation, whether within a single response agency or among several response agencies, is critical to a successful operation. A person or agency that takes it on itself to carry out actions that are not coordinated through a structured **incident command system** (ICS) is called a *freelancer*. Freelancing at an emergency scene disrupts the efficiency of the operation, endangers other responders, and leads only to failure of the entire operation.

incident command system (ICS)
a standardized on-scene emergency management system designed to provide an integrated organizational structure that reflects the complexity and demands of single or multiple incidents, without being hindered by jurisdictional boundaries

Most large-scale events require that different tasks and objectives be assigned to specified groups, branches, or divisions typically referred to as sector leads. Each of these *sector leads* reports back to the incident commander after the task is completed or when further overall direction is required. Incident commanders must give their sector leads the autonomy to perform the mission assigned to them. When assistance from another sector is required, however, this request should be placed through the incident commander, so that the IC can direct resources as needed to ensure that all responders are working toward the same goal.

UNIFIED COMMAND

unified command structure
an application of ICS used when there is more than one agency with incident jurisdiction at a scene, or during an incident with cross-political jurisdictions

All emergency response agency personnel must be trained to function within a **unified command structure**. Some states require formalized training in their emergency planning structure, which is compliant with NIMS. An example is the Wisconsin Emergency Management Agency and its WERP (Wisconsin Emergency Response Plan). The WERP serves as a coordinating document for supporting plans, such as individual agency plans (IAPs); federal, state, and local government plans; and private organization plans. The names of such plans may differ slightly from state to state or region to region, but all plans are alike in concept and principle, and conform to NIMS standards.

Most agencies have preplanned lists of response contingency plans for various types of emergencies, including aircraft accidents. In addition, different agencies generally focus on certain tasks. For example, once ARFF teams arrive on the scene, they will concentrate on extinguishing the aircraft fire and performing rescue. In a military aircraft accident, military medical personnel (called bioenvironmental personnel) are tasked with determining whether there is a contamination hazard. (These responders coordinate with civilian medical authorities, however.) In a civilian aircraft crash, local medical authorities handle this job, although they may call on military medical resources in the event of a mass casualty incident. The responsibility of safeguarding military classified materials is relegated to military security authorities.

National Transportation Safety Board (NTSB) federal agency charged with investigating and determining the reason for and contributing factors to aircraft, railroad, maritime, highway, and pipeline transportation accidents; makes recommendations to facilitate the prevention of additional mishaps or accidents

If the **National Transportation Safety Board (NTSB)** is called to the scene, however, local agencies should brief the NTSB personnel and then relinquish authority to them. (If this occurs, however, you should remain on site as needed, working as part of the unified command and providing continuing firefighting, rescue, HAZMAT, and other related duties.) The NTSB is an independent federal agency responsible for highway, aviation, marine, railroad, and pipeline safety. It is charged with investigating transportation accidents and determining the most likely cause(s) of these accidents. After studying the causes, NTSB issues safety recommendations. The NTSB also studies transportation safety issues and evaluates the effectiveness of government agencies involved in transportation safety. The *Canadian Transportation Accident Investigation and Safety Board* performs the same functional and regulatory duties in Canada as the NTSB does in the United States. It has been proven that military and civilian authorities can effectively implement a good unified command operation, which includes NIMS.

If criminal or terrorist activity is suspected in an aircraft incident, the FBI becomes involved in the investigation and becomes part of the incident unified command structure.

What Agencies Are Needed?

The first and foremost goal of any rescue operation is to save lives and protect property. As previously noted, it may be necessary for your agency to request help from a variety of other response agencies to achieve this goal. Determine which agencies in your community can offer some form of help in the event of a large-scale event. If an aircraft goes down in water, for instance, boats will be needed. In a city or municipal setting, falling debris and or fire from the accident may damage several structures. You may need additional pumpers, hose tenders, aerial ladder apparatus, and personnel to combat adjacent structural exposures. In populated areas, large amounts of smoke may necessitate that citizens in the downwind path of the smoke be evacuated. Until an NTSB Accident Investigation Team arrives, allow only the following personnel on scene:

- FAA
- Police (or other law enforcement)
- Fire
- Medical Examiner/Coroner
- Airport fire department, if available

In addition to these agencies, there may be state or federal agencies with assets you may need to gain control of a large scale accident.

Roles of Agencies within the Command Structure

Coordination among agencies can be difficult, at times. It is the incident commander's responsibility, however, to bring the separate agencies into a single cohesive team that has a common goal, has clearly understood roles and duties, and is "speaking the same language."

Every response agency that is called to the scene of an aircraft accident has a role to fulfill in order to help manage the incident. Fire departments are tasked with protecting life and property; paramedics with patient care; law enforcement with scene control and preservation of evidence; NTSB with investigation; and so on. Each team leader has sub-goals in the context of the overall emergency response. It is the responsibility of the incident commander to understand each agency's role and sub-goals and to coordinate these with the overall goals of incident command. Personnel from all agencies must work together to establish a foundation early in the incident to effect a positive outcome. If tasked as the incident commander, you may want to make a portable display board that illustrates how each agency's key personnel fill the various leadership positions of a multi-agency rescue operation. The key personnel from other agencies are specialists in their respective fields, and representing them in the unified command structure is a necessity for operation success, not a diplomatic gesture. When conducting command briefings, ensure that everyone understands what the mission goals are, then allow them to carry out their individual roles as part of the overall plan of action. Once everyone's goals and roles are outlined, let everyone do their respective jobs. As long as the various aspects of the scene are being properly managed, there is no need to micromanage them.

WHAT'S YOUR PLAN?

Any serious emergency poses a host of challenges that can overwhelm the best incident commanders, company officers, and frontline rescuers. To manage a large-scale incident, such as an aircraft accident, you must have a plan. When creating your plan, use the simple acronym *PORT* to help you divide each tactical, command, or logistical problem into manageable segments, step by step:

P *Problem:* Identify one at a time.

O *Objective:* How can you solve the problem?

R *Resources:* What resources do you have available to fix the problem?

T *Time:* How much time is required to fix the problem?

Similarly, the *KISS principle* is a good memory jogger for developing tactics and procedures in this potentially complex environment.

K *Keep*

I *It*

S *Safe* and

S *Simple*

The amount of resources needed to deal with an aircraft accident or any mass casualty incident is based on the total injuries, the aircraft's size, the extent of property damage, and so on. Preplan for such events by determining what equipment may be needed from other resources, where large vehicles can gain entry to the scene, and a command post location for base-camp operations. In addition, you should determine where you can stage responders downstream if required to fight flowing fires, reach victims, stop the spread of hazardous materials, and/or protect evidence.

If an aircraft accident includes **mass casualty incident (MCI)**, notify the appropriate agencies and personnel, and follow your **emergency response plan (ERP)** (or your local emergency response plan [LERP]) for such events. An MCI is an event that results in many patients suffering severe trauma. In such situations, for instance, you may have to request helicopters for transporting severely injured people to trauma centers. Follow the SOG that you would follow for any other aeromedical evacuation.

mass casualty incident (MCI)
any accident or catastrophe that involves large numbers of casualties

emergency response plan (ERP)
maintained by various jurisdictional levels for managing a wide variety of potential hazards; also, local emergency response plan (LERP)

Your Existing Standard Operating Guidelines

As you develop your Standard Operating Guidelines for aircraft accidents, take into consideration the types of fire equipment your department has and whether they have foam-delivering capability. You need to know the amount of foam concentrate this equipment carries and the delivery rate of the foam agent (foam-water mix) each apparatus can achieve.

In addition to your own equipment, know what types of equipment nearby fire departments are willing to send, including aircraft rescue firefighting (ARFF) apparatus. Even if nearby departments don't have ARFF equipment, other equipment may be useful. Although it is preferable to use Class B firefighting foam at aircraft accidents, in some incidents, fuel-fed fires have been successfully extinguished using high-pressure water fog with dry chemical fire extinguishers.

You also need to list the rated pumping capabilities of the responding equipment, the radio call signs of these response vehicles, and their projected response time. Your SOG should contain a list of tactical radio frequencies enabling all responding agencies to be prepared for efficient communications.

Crashes involving large-frame aircraft may ignite a large fire, because these airplanes may carry thousands of gallons of fuel. Your Standard Operating Guidelines should indicate what and how much equipment is needed to suppress such fires, including a primary and secondary means of delivering foam concentrate to the scene. Ensure that your response plans address the control of large fuel fires in the event that there is not an adequate foam supply close to the incident. Extra foam concentrate may be obtained from a nearby airport, maritime shipping port, petroleum storage facility or refinery, or military installation. The SOG should specify the sources of available foam, as well as the names and phone numbers of source contacts and alternates. Your communications center must know how to contact any of these people at any time. As part of your preplanning, you may have to prearrange for a trucking company to transport raw foam to an emergency scene, which may require a police escort.

■ Note
Ensure that your response plans address the control of large fuel fires in the event that there is not an adequate foam supply close to the incident.

Many aircraft fires are large and likely to generate mass casualties. Your SOG for aircraft accidents may require other specialized equipment, such as lighting equipment for night operations. The source of this equipment should be listed in your LERP, as should sources of additional equipment and agencies, including medical authorities, the coroner's office, police, critical-incident stress debriefing teams, the health department, and environmental management personnel. The nearest military installation may have resources not available at all, or in sufficient quantity, from your local government. These may include explosive ordinance disposal personnel, medical personnel, law enforcement, hazardous materials personnel, bioenvironmental personnel, heavy equipment, a vehicle fuel, emergency generators, emergency lighting, emergency heating equipment, emergency cooling equipment, tents, additional manpower, and many other resources.

Your Existing Standard Emergency Management System

Your emergency management system may have characteristics that make it unique from other systems being used in your community and by other responding agencies. It is important for all responding agencies to understand the components of the given system being used at an accident site. The best way to achieve this interagency plan familiarity is for agencies to train with one another to identify and correct flaws in existing systems. This way, all responders can make a unified effort to ensure that any emergency operation runs smoothly and effectively. When adjustments to a program are made, ensure that all personnel, within and outside of your agency, have the opportunity to understand and practice these procedural changes *before* an incident occurs. Practice does not eliminate every confusing issue and potential mistake, however, it minimizes major gaps in command plans. When you develop a plan, don't let it sit and collect dust—train with it, practice it, evaluate it, and adjust it until the process works for you and your community. When you get it right, start the training, practice, and evaluation all over again. The more time you spend refining your existing standard emergency management system, the stronger it becomes. When developing your plans, consider the role that weather plays in them. Excessive heat, snow, or rain will affect all aspects of your rescue and firefighting operation.

■ Note **The more time you spend refining your existing standard emergency management system, the stronger it becomes.**

You can also improve your local emergency response plans by studying plans from agencies that are similar to yours. These plans can be easily accessed on the Internet. Again, although the names of these plans may differ, all of the plans conform to NIMS standards.

The Impact of Terrorism on Your Plans

weapons of mass destruction (WMD) (1) any incendiary, poison gas, bomb, grenade, rocket having a propellant charge of more than 4 ounces or missile having an explosive or incendiary charge of more than 0.25 ounce, mine, or similar device; (2) any weapon that is designed or intended to cause death or serious bodily injury through the release, dissemination, or impact of toxic or poisonous chemicals or their precursors; (3) any weapon involving a disease organism; or (4) any weapon that is designed to release radiation or radioactivity at a level dangerous to human life

Aircraft are inviting targets for terrorist activities because they carry a high concentration of people and because extensive collateral damage may result from an act of terror involving an airplane. This fact was brought forcibly to the awareness of the world on September 11, 2001, when terrorists flew planes into the towers of the World Trade Center and into the Pentagon. The threat of deliberate terrorist acts on aircraft is still present. In the summer of 2004, terrorists detonated bombs that resulted in the crashes, minutes apart, of two passenger airliners in the skies over Russia. In 2006, in England, a plot to place bombs on several passenger-carrying airliners bound to the U.S. was foiled. Throughout the civilized world, concerns about aviation terrorism are shared by the airlines, military forces, law enforcement, fire departments, disaster support agencies, disaster planners, aircraft pilots, and the people who pay to travel on airplanes. Although flying is still the safest mode of transportation, there is genuine concern about air-related terrorism, and emergency response agencies at all levels are conducting additional training for managing aviation disasters associated with terrorism.

In compliance with the Department of Homeland Security, you are encouraged to review your response role as it applies to the NIMS command/disaster management system, as well as any local terrorism response procedures and responsibilities (see **Figures 4-1** and **4-2**). Become familiar with all NIMS and terrorism terminology, such as **weapons of mass destruction (WMD)**.

The National Incident Management System

As noted earlier, when two or more agencies respond to the same emergency, a common ground for incident management must be formed. On February 28, 2003,

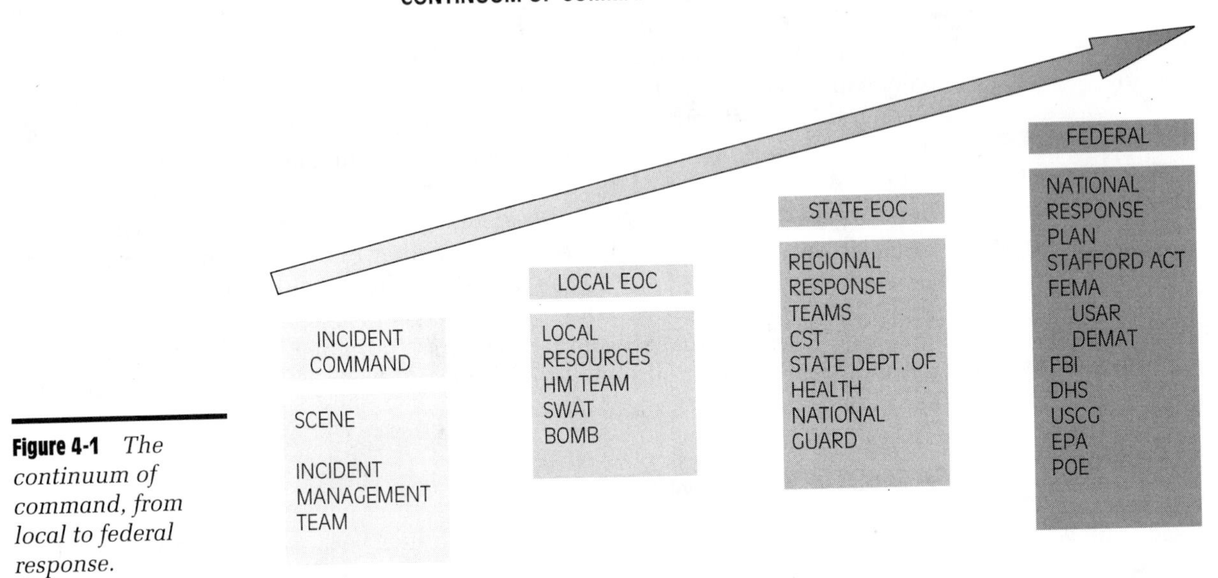

Figure 4-1 *The continuum of command, from local to federal response.*

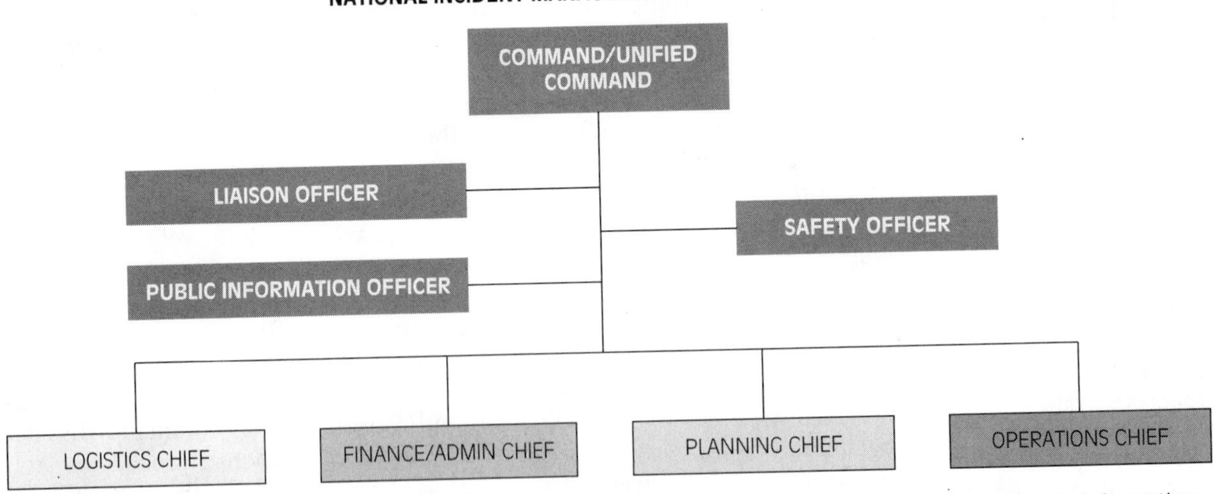

Command is comprised of the incident commander and the command staff. Command staff includes the public information officer, safety officer, and liaison officer as assigned by the incident commander.

Command can also include a number of commanders from different legal, geographic, and functional responsibilities working together making decisions for the incident.

General Staff is comprised of four major functional elements and are headed up by operations section chief, planning section chief, logistics section chief, and finance/administration section chief.

Figure 4-2 *The National Incident Management Structure.*

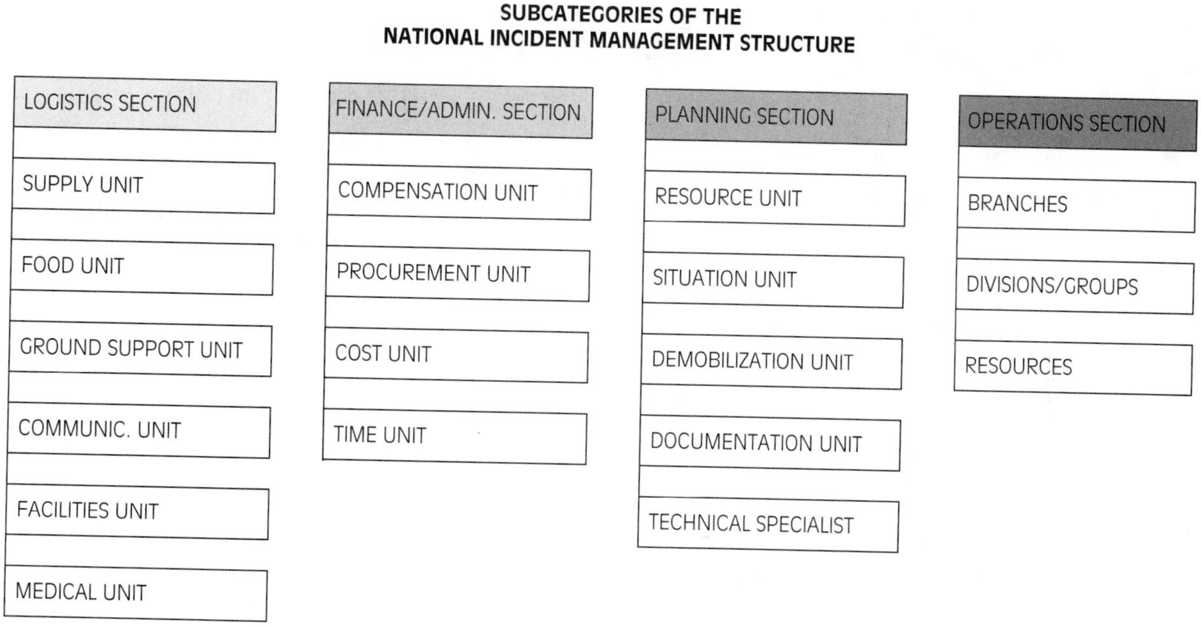

SUBCATEGORIES OF THE
NATIONAL INCIDENT MANAGEMENT STRUCTURE

LOGISTICS SECTION	FINANCE/ADMIN. SECTION	PLANNING SECTION	OPERATIONS SECTION
SUPPLY UNIT	COMPENSATION UNIT	RESOURCE UNIT	BRANCHES
FOOD UNIT	PROCUREMENT UNIT	SITUATION UNIT	DIVISIONS/GROUPS
GROUND SUPPORT UNIT	COST UNIT	DEMOBILIZATION UNIT	RESOURCES
COMMUNIC. UNIT	TIME UNIT	DOCUMENTATION UNIT	
FACILITIES UNIT		TECHNICAL SPECIALIST	
MEDICAL UNIT			

Figure 4-3 *National Incident Management Structure subcategories and units.*

President Bush issued Homeland Security Presidential Directive-5. HSPD-5 directed the Secretary of Homeland Security to develop and administer the National Incident Management System (NIMS). NIMS provides a consistent nationwide template to enable all government, private-sector, and nongovernmental organizations to work together during domestic incidents.

Figure 4-3 shows the assorted branches of the NIMS command structure. NIMS is revised and updated as additional needs, procedures, and dangers are discovered. Emergency responders are encouraged to access the NIMS Web site often in order to stay apprised of changes to this program. Updated information is available via publications, bulletins, and, as mentioned, the Internet.

federal emergency management agency (FEMA)
part of the Department of Homeland Security; intended to create a more efficient system to help America prepare for, respond to, and recover from all forms of disasters

The **Federal Emergency Management Agency (FEMA)** Independent Study Program IS-700 National Incident Management System (NIMS) is available at http://www.training.fema.gov. This course should be completed in conjunction with IS-100 Introduction to Incident Command System. ICS-100, Introduction to the Incident Command System, provides the foundation for higher-level ICS training. This course describes the history, features, principles, and organizational structure of the ICS, and explains the relationship between ICS and NIMS.

SUMMARY

This chapter touched on several areas to consider during response to an aircraft accident.

- Knowing your response procedures is crucial for a successful outcome.
- We as emergency responders need to know the fastest response routes as well as alternate routes to any accident site. This means knowing as many aspects of your response area as possible.
- By regular training, and preplanning, we will be better prepared for what to expect when we arrive. This should include knowing where to access stockpiles of foam concentrate, specialized rescue equipment, and trained specialized, auxiliary, and private sector personnel any time, any day.
- Responses to waterways can take on a new life because of the challenges posed by gaining access to an accident in a lake, river, or the ocean.
- We discussed the importance of someone taking command and being the sole source of direction. A unified command system is the best means to link different agencies so they can work together towards a common goal: Managing the Incident.
- Finally, response plans were discussed: Your plan is your plan. You should train all personnel who use, practice, and evaluate it. If it is discovered that your plan is becoming obsolete due to changing circumstances in your community, address the changes and update your plan.
- Elements on terrorism and the presidential direction to implement a common National Incident Management System were touched. Be vigilant for updates and changes in various aspects of the MIMS System. It is up to you to ensure your agency is able to perform as expected within the NIMS Structure.
- Chapter 5 covers several response tactics and strategies.

KEY TERMS

Emergency response plan (ERP) (Also, local emergency response plan [LERP]) The plan maintained by various jurisdictional levels for managing a wide variety of potential hazards.

Federal Emergency Management Agency (FEMA) Now part of the Department of Homeland Security, FEMA is intended to create a more efficient system to help America prepare for, respond to, and recover from all forms disasters.

Incident command This determines the success or failure of an emergency response. Incident command is especially critical during aircraft accident rescue, because crew and passengers are encased in these craft and surrounded by large volumes of flammable aviation fuel, as well as baggage and cargo that may contain hazardous materials.

Incident command system (ICS) A standardized on-scene emergency management system designed to provide an integrated organizational structure that reflects the complexity and demands of single or multiple incidents, without being hindered by jurisdictional boundaries. ICS is the combination of facilities, equipment, personnel, procedures, and communications operating with a common organizational structure to aid in the management of resources during incidents. ICS is used for all kinds of emergencies, and it is designed to be in compliance with NIMS.

Incident commander (IC) The person who is designated as being responsible for all activities associated with an incident, including the development of tactical plans, strategies, and the utilization of resources. The IC has overall authority and responsibility for managing incident operations and is responsible for the direction of all incident operations at the incident site. He/she also has the authority to demobilize response personnel, equipment and other resources as the situation reduces in size, magnitude, or degree of hazard.

Mass casualty incident (MCI) Any accident or catastrophe that involves large numbers of casualties.

National Transportation Safety Board (NTSB) The federal agency charged with investigating and determining the reason for and contributing factors to aircraft, railroad, maritime, highway, and pipeline transportation accidents. This agency makes recommendations to facilitate the prevention of additional mishaps or accidents.

Unified command structure An application of **ICS** used when there is more than one agency with incident jurisdiction at a scene, or during an incident with cross-political jurisdictions. Each agency designates an incident commander at a single *incident command post (ICP)*. The agencies work through these designated members of the unified command to establish a common set of objectives and strategies and a single *incident action plan (IAP)*.

Weapons of mass destruction (WMD) (1) Any incendiary, poison gas, bomb, grenade, rocket having a propellant charge of more than 4 ounces or missile having an explosive or incendiary charge of more than 0.25 ounce, mine, or similar device; (2) any weapon that is designed or intended to cause death or serious bodily injury through the release, dissemination, or impact of toxic or poisonous chemicals or their precursors; (3) any weapon involving a disease organism; or (4) any weapon that is designed to release radiation or radioactivity at a level dangerous to human life.

REVIEW QUESTIONS

1. What does “HE” stand for?
2. What are several items classified as weapons of mass destruction?
3. The simple PORT concept allows each tactical objective to be separated into segments. What does the acronym stand for?
4. What factors are involved in preplanning with potential responding agencies?
5. In your own words, what is NIMS and its function?
6. In your own words, what is the NTSB and its function?
7. Because GPS satellite mapping systems may malfunction, what contingency should you have to locate the area where a plane crash has been reported?
8. What is a freelancer?
9. What are the problems that result from freelancing?

Figure 4-4 *Aircraft disaster training.*

10. Study the images in **Figure 4-4.** This aircraft disaster drill was conducted on a hot, dry afternoon.

The drill site is at a location that is below sea level. To the right of the picture is a large river. Note the location of the elevated railroad trestle and the elevated highway.

To the left (not visible in the picture) is a large residential community with extremely narrow streets. The fire hydrants are rated at 500 GPM. Note the depth of tire tracks on the dry grass. How would your plan address the following concerns?

- Rail traffic downwind from the smoke
- Highway traffic downwind from the smoke
- The possibility of looters and sightseers from the nearby neighborhood
- Excessive rain
- Water supply concerns

11. What is your role as it applies to the NIMS command/disaster management system?

12. What are several resources you could obtain from a nearby military base to help you handle an aircraft accident?

STUDENT EXERCISES

Typical Airport Emergency Action Plan

Review the Table of Contents in **Figure 4-5.** Research and compare with your local ERP. Where are there similarities? Where are there differences? Become familiar with your local ERP.

HAZARD SPECIFIC APPENDICES

Figure 4-5 *Cover sheet/index for a major international airport.*

RESPONSE TACTICS AND STRATEGIES

Learning Objectives

Upon completion of this chapter, you should be able to:

- Understand the grim realities—including the multitude of sights, sounds, and smells—you may encounter at the scene of a large-scale aircraft crash, as well the need to prepare for this emotionally overwhelming information and how to deal with it.
- Understand how to safely and precisely approach a downed aircraft.
- Understand tactical considerations for the following conditions:
  - crashes of various impact levels (high, medium, and low)
  - crashes on hard surfaces and difficult terrain
- Understand basic information about fire-extinguishing agents:
  - water fog/spray
  - Class B firefighting foams
  - dry chemical
  - dry powder
- Understand how to assist airport fire departments and other responders in the following activities:
  - firefighting
  - rescue and removal of passengers
  - resupply

- Understand how, while assisting an airport fire department, to effectively respond to mass casualty incidents:
  - protecting yourself
  - basic triage
- Understand common helicopter operations, including aeromedical evacuations (medevac).
- Develop strategies for planning ahead, including:
  - developing a tactical checklist
  - engaging in airport emergency exercises
  - caring for your crew.

INTRODUCTION

An aircraft crash is a terrible and complex situation to manage. **Figure 5-1** shows a typical example of what such an event can look like as you approach the scene from a distance. Responding to an aircraft incident requires that many questions be considered: Is it a passenger plane with many people on board? Is it a military aircraft that may be carrying weapons and munitions? What about water supply? Is there enough foam nearby? Such concerns can be countless. The incident

Figure 5-1 *Approaching an aircraft incident.*

might be fairly simple, for example, involving a small private plane that has crash-landed in a vacant field and whose passengers have walked away with minor injuries. Conversely, an incident may involve a large-frame aircraft bearing thousands of gallons of fuel and many passengers. It might be a military or commercial cargo aircraft. If it is a military aircraft, weapons may be on board. Often, an aircraft incident taxes most, or even all, of a community's response resources, including fire departments, law enforcement, hospitals/EMS, and Red Cross and other organizations.

THE GRIM REALITIES OF A CRASH SCENE

You rarely have the opportunity to determine response objectives and tactical considerations before responding to an aircraft crash.

Nothing can fully prepare you for the experience of dealing with an aircraft accident, especially a large-scale, mass-casualty incident, for the first time. An aircraft crash site is a nightmarish scene whose sheer chaos, enormous number of dangers, and widespread destruction can be stunning.

Thus, your preplanning should include information about what you are likely to face during such an event, so that you have some idea of what to expect and can prepare for it mentally, at least to some degree. For example, try picturing the largest structure fire or worst automobile accident you have responded to, and multiply it by ten. Such exercise is critical, because unless you prepare for the shock of seeing such devastation by reading descriptions of it, visualizing it, and planning how you will respond to it, the shock may interfere with your ability to function effectively. *Remember:* You are the answer to someone's cry for help—if you are unable to focus on tactical objectives and strategic goals, you may do more harm than good.

The first challenge you face when responding to an aircraft incident is the initial shock of seeing the scene. If the incident involves a large-frame aircraft, the sights can include a huge debris field and large fires. Aircraft wreckage, personal effects, cargo, and human remains may be scattered over a wide area. This is especially true in the event of a mid-air collision or an aircraft's in-flight breakup due to explosion or other catastrophic disintegration of the aircraft's structure. The unnerving roaring of fire out-of-control, the screams of injured people, the sound of pressurized containers exploding, and the shouts of bystanders fuel the chaos. In addition, you may be momentarily overcome by the smells of burning or spilled fuel or hydraulic fluids, of burning plastics, metals, and cargo. If people have died during the crash, you are likely to encounter the horrific odors generated by burned or decaying bodies.

Managing the Stress

When you feel yourself being overwhelmed, emotionally and/or physically, by the tragedy and confusion surrounding you during a crash situation, take a deep breath and concentrate on the end result you are working toward, and then proceed. To help you remain focused and able to tackle the myriad problems you face, remember and apply the PORT acronym discussed in Chapter 4: Quickly and simply define each objective, and then address each concern using your incident command and unified command procedures. If you are the first-in fire-suppression unit (i.e., the first to arrive on scene), you may be able to pass command to the next-in unit. This may be necessary when resources are limited and immediate tactical actions are required to save lives.

■ Note
Smell is a powerful memory trigger: It can raise terrible memories of an incident even years after the responder has experienced the traumatic event.

Smell is a powerful memory trigger: It can raise terrible memories of an incident even *years* after the responder has experienced the traumatic event. Worse, smell can trigger incidences of post-traumatic stress disorder (PTSD), a severe form of depression related to a traumatic event. (PTSD is discussed in detail in Chapter 6.)

Several methods exist to minimize your exposure to the smells generated by an aircraft accident. *The suggestions provided here have proven to be effective at actual incidents.* If you are working in an area where self-contained breathing apparatus is required, the SCBA will limit or even eliminate your exposure to smells. If an SCBA is not needed, dab small amounts of mentholated chest rub or mint-flavored toothpaste on the inside of a dust mask, then wear the mask to minimize the intensity of odors. If tasked with retrieving human remains, wear latex, nitrile, or vinyl gloves beneath thick leather work gloves. This reduces your sense of touch somewhat, thereby minimizing the intensity of the unpleasant task. (It also reduces your exposure to diseases, which is discussed later in the chapter.)

APPROACHING AIRCRAFT ACCIDENTS

Whether an accident happens on a country road, on an airport runway, in a business district, or in a residential neighborhood, you must know how to approach a downed aircraft accident safely, because, as previously discussed, the fire resulting from a crash requires rapid control and extinguishment. The best tactic to extinguish a fire involving large quantities of fuel is mass application of a **fire-extinguishing agent**, such as Class B foam, water fog, or a combined dry chemical-water application, which can be started on the initial approach.

fire-extinguishing agent
material used to extinguish a fire, such as Class B foam, Class D dry powder, water fog, hose streams, gaseous agents (e.g., carbon dioxide or Halon), or a combined dry chemical-water application

Safe Approach

During your approach to a scene, be alert for indications of danger, survivors, and clues to the cause of the accident. It is crucial to exercise caution when approaching a crash site, because the bodies of deceased persons and of survivors may be scattered about the area. Many air-crash survivors have been injured either by the post-crash fire or by being struck from incoming emergency vehicles. As mentioned in Chapter 4, driving a vehicle quickly through smoke can endanger victims, bystanders, and other rescuers.

For example, you may encounter crewmembers that have safely ejected from the stricken aircraft. These people may have suffered fractures or neck injuries from the intense forces of the ejection seat and the accompanying blast of wind, and they may be unable to move out of the way of an approaching vehicle.

Another consequence of approaching a site too fast and without first thoroughly surveying the scene is that you may drive your vehicle into a ditch or similar obstacle, rendering it useless. Also, as you bring your vehicle closer to the aircraft, watch for and avoid externally mounted fuel tanks and munitions—these may still be attached to the aircraft, or they may have been torn from it and be scattered about the crash area.

TACTICAL CONSIDERATIONS

National Defense Area (NDA)
an area established on nonfederal lands located within the United States or its possessions or territories, for the purpose of safeguarding classified defense information or protecting Department of Defense (DOD) equipment and/or material by keeping non-authorized people away from such sensitive materials

! Danger
Approach military fighter aircraft at a 45-degree angle from the forward side of the fuselage to avoid missiles, machine guns, or cannons.

Approaching an aircraft accident scene is the first step to gaining control of a chaotic situation. Approach an incident site from an *upwind* (i.e., the wind is at your back) and uphill vantage whenever possible; doing so gives you the best view of the scene and help you visually locate spilled fuel and hazardous materials, so that you can avoid them as you bring your apparatus onto the site grounds.

As you scan the incident site from your approach vantage, determine the easiest point of access for your apparatus, as well as the best placement of it. *Vehicle positioning* is critical, because you must find a place to park and position your vehicle from which it will not have to be moved or, if it does, then moved only minimally. Also, the location of an aircraft's main entry/exit door is important to vehicle placement, because it's vital to ensure that the emergency egress path from the aircraft to emergency vehicles is not blocked. The main door on most airplanes that carry freight and passengers is located on the left side of the aircraft as it is facing forward. The main door of smaller, general aviation aircraft, such as the Cessna 172, is located on the aircraft's right side.

If the crash involves an armed combat aircraft, your approach and vehicle positioning may have to be adjusted to avoid weapons. **Figure 5-2** shows the danger areas posed to responders by a typical combat military aircraft.

Approach military fighter aircraft at a 45-degree angle from the forward side of the fuselage to avoid missiles, machine guns, or cannons. Missiles pose danger both directly in front of and behind a downed combat aircraft.

Once your apparatus is positioned, accessing and rescuing any personnel trapped in the aircraft is your first concern.

Then, try to determine whether the downed aircraft is a private, commercial passenger, commercial cargo, or military aircraft. If the airplane is military, you may be requested to establish a **National Defense Area (NDA)**. An NDA is a site

AIRCRAFT HAZARDS

1. ARMAMENT FWD FIRE ZONE - 1000 FT.
2. RADAR - 300 FT. PERSONNEL - 500 FT.
3. ENGINE AIR INTAKES - 25 FT.

CAUTION

DANGER ZONE CAN EXTEND AS FAR AS 5 FEET AFT OF THE AIR INLET AT HIGH POWER SETTINGS.

4. CANOPY JETTISON ENVELOPE - 50 FT.
5. JET FUEL STARTER (JFS) INTAKE - 4 FT.
6. JFS EXHAUST AT IDLE OR ENGAGEMENT - AFT ALONG THE CENTERLINE TO THE ENGINE TAIL CONES. TEMPERATURE: 1000 +/− 180 DEGREES. RMP DOES NOT MATTER.
7. TURBINE BLADE FAILURE - 300 FT.
8. ENGINE EXHAUST -
   - IDLE RPM 0 - 25 FT: 200 DEGREES
   - AT 80% 0 - 40 FT: 200 DEGREES
   - AT MIL 0 - 15 FT: 800 DEGREES
   - AT MAX 0 - 20 FT: 3000 DEGREES
   - VELOCITY: ABOVE 1000 MPH AT THE TAILPIPE.
9. ARMAMENT EXHAUST

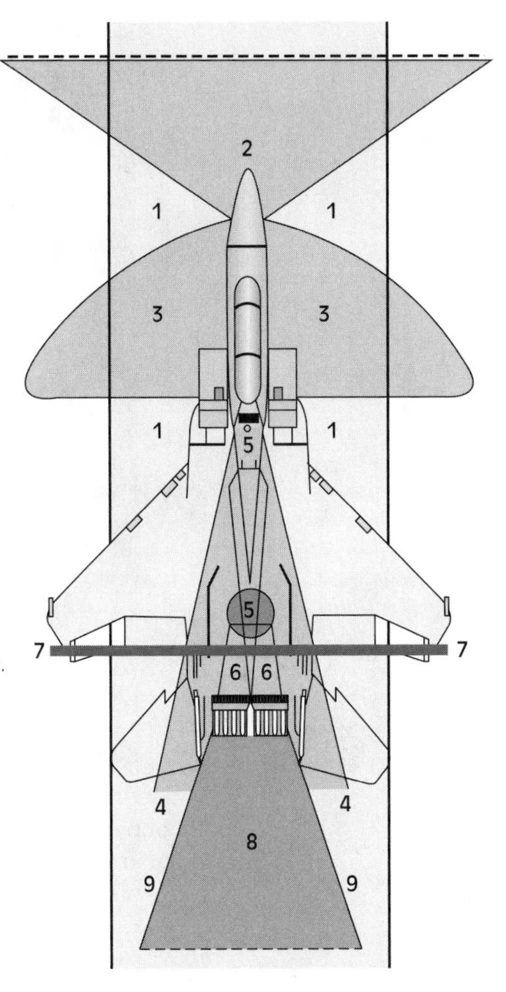

Figure 5-2 *Danger zones around a typical fighter aircraft. (Courtesy of the U.S. Air Force.)*

established on nonfederal lands located within the United States or its possessions or territories for the purpose of safeguarding classified defense information or protecting Department of Defense (DOD) equipment and/or material. When a site is declared to be a National Defense Area, it is temporarily placed under the effective control of the DOD. An NDA should be declared only as a result of an emergency event. The senior DOD representative at the scene defines the area's boundaries, marks them with a physical barrier, and posts warning signs. The landowner's consent and cooperation is obtained whenever possible; however, military necessity dictates the final decision regarding location, shape, and size of the national defense area.

As soon as practical, once the initial attack has been completed and trapped crew and passengers have been rescued, draft a basic site safety escape plan for

Figure 5-3 *Provide hose streams to protect rescue personnel.*

rescuers and those being rescued. It is crucial to identify an expedient way out, or an **emergency escape route**, in the event that conditions change and place rescue personnel and survivors in danger.

It is also important to approach a crash site from an upwind position to avoid fires and other dangers. The large fuel-fed fires that commonly result from aircraft crashes produce enormous amounts of heat and smoke. In addition, if a site's debris field is large, it may generate multiple fires, and fire impingements may threaten surrounding structures, apparatus, or objects. In such cases, the incident commander may need to request additional water-tender fire trucks to shuttle water to firefighting vehicles.

■ Note
Streams of firefighting-agent should be directed to cover rescuers who are entering and occupants who are exiting a burning airplane.

Streams of firefighting-agent should be directed to cover rescuers who are entering and occupants who are exiting a burning airplane. Notice in **Figure 5-3** that the firefighter performing rescue is protected by hoselines and additional personnel.

Figure 5-4 shows the fundamental **tactics** for dealing with a low-impact crash landing involving a small business jet. Notice the location(s) of incident command and the positioning of firefighting, rescue, and water-supply vehicles. The figure shows police officers first on scene and attempting quick access to the people inside the aircraft. (Case studies have shown that police are most often the first on scene.) Fire vehicles have been positioned and hoselines have been

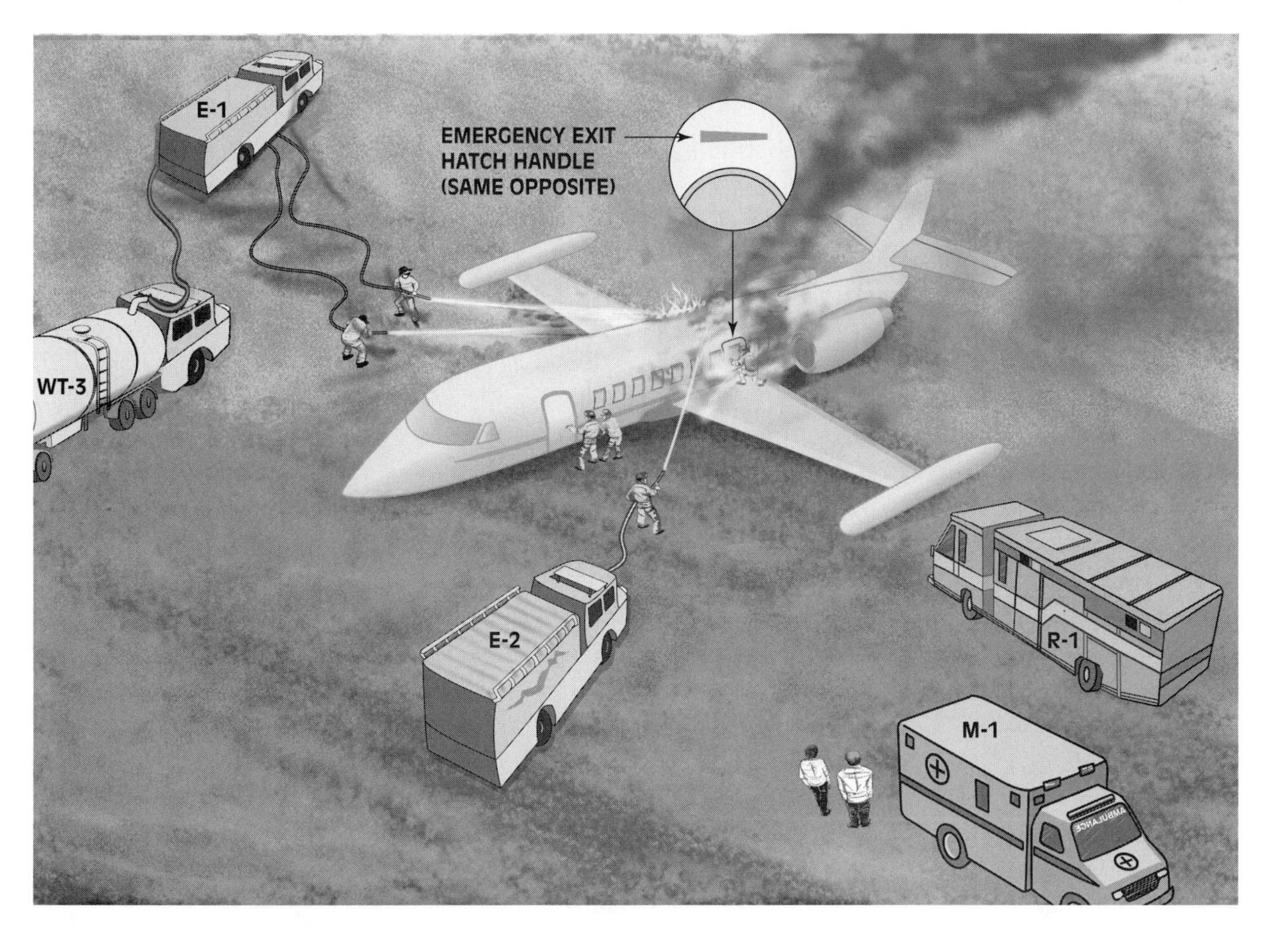

TACTICAL DIAGRAM BUSINESS JET CRASH LANDING

E-1—Has 600 gallons of water and 50 gallons AFFF foam concentrate. Begins initial attack, using two 1 1/2" pre-connect hoses, interlocking fog water spray patterns. It is possible to contain a fuel-fed aircraft fire in this manner. Being resupplied by a 2,500 gallon Water Tender.

E-2—Has advanced one pre-connect 1 1/2" pre-connect hose to cover rescue. Has 600 gallon tank, will be resupplied by next in Water Tender.

WT-3—2,500 Water Tender resupplying E-2. (often called a "Nursing Operation")

R-1—Rescue Squad, performs rescue operations/entry while coverage being provided by other units. Their entry into the aircraft backed up by hose line people.

M-1—Establishes/monitors Triage Area.

INCIDENT COMMAND—Is set up at area affording safety and visibility.
Additional resources such as AFFF foam, and manpower available on request.

Figure 5-4 *A private jet making an off-airport crash landing.*

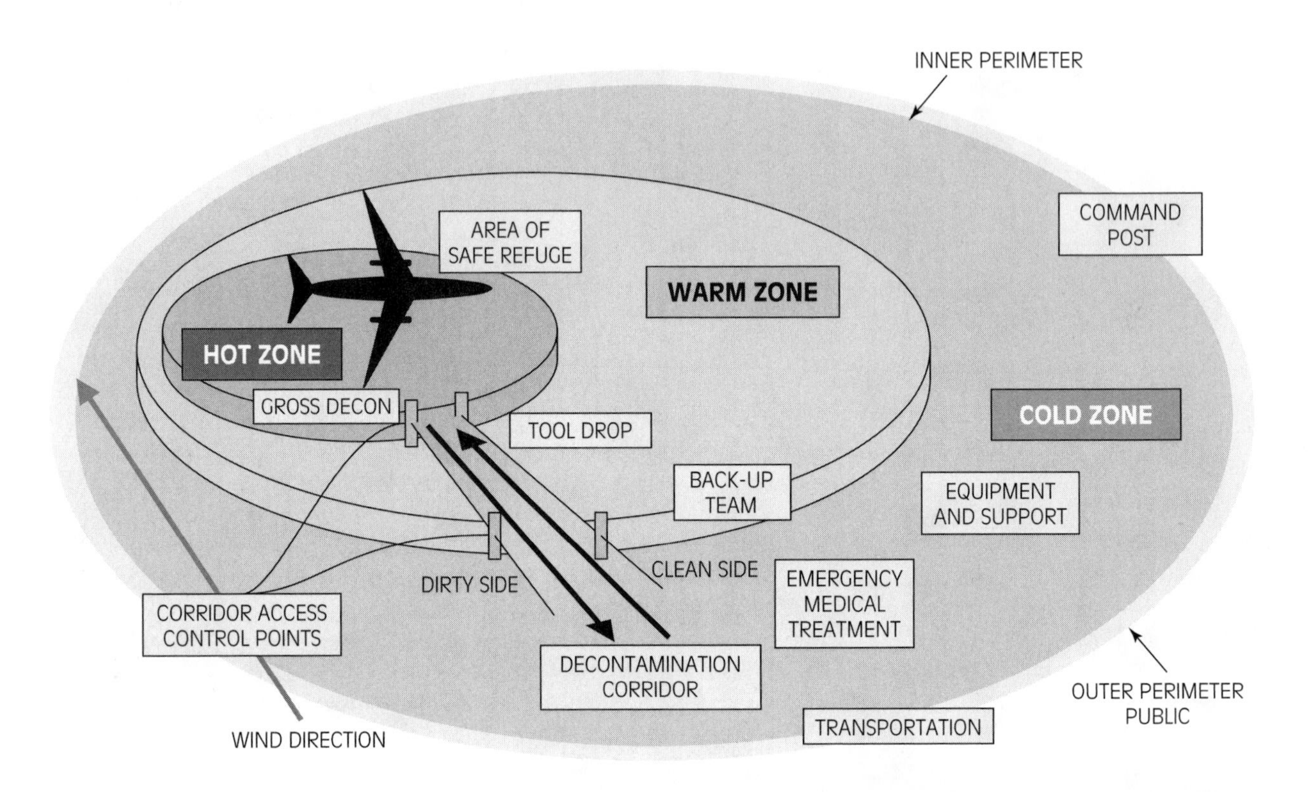

Figure 5-5 *The zones for a major aircraft accident or a HAZMAT.*

• Caution
Water may be used on a wheel fire to protect adjacent portions of the airplane and cool the tires. Never, however, apply water to burning magnesium.

advanced to extinguish the fire and protect rescuers and escaping passengers. In this scenario, ARFF vehicles are en route, as are county fire vehicles.

Water may be used on a wheel fire to protect adjacent portions of the airplane and cool the tires. *Never*, however, apply water to burning magnesium.

Figure 5-5 is a diagram of a generic major aircraft accident or a hazardous materials operation. Notice that provision is made for a refuge area close to the hot zone.

Figures 5-6 and **5-7** show the proper methods for establishing an effective *decontamination corridor* for a hazardous materials incident. After any major aircraft fire (especially if composite materials are involved), it is advisable, at the minimum, to perform a *gross decontamination* at the scene. Then, properly launder any turnout clothing suspected of having been exposed to large amounts of toxic smoke or hazardous chemicals according to applicable policies and regulations. Currently, these are found in NFPA Standard 1500.

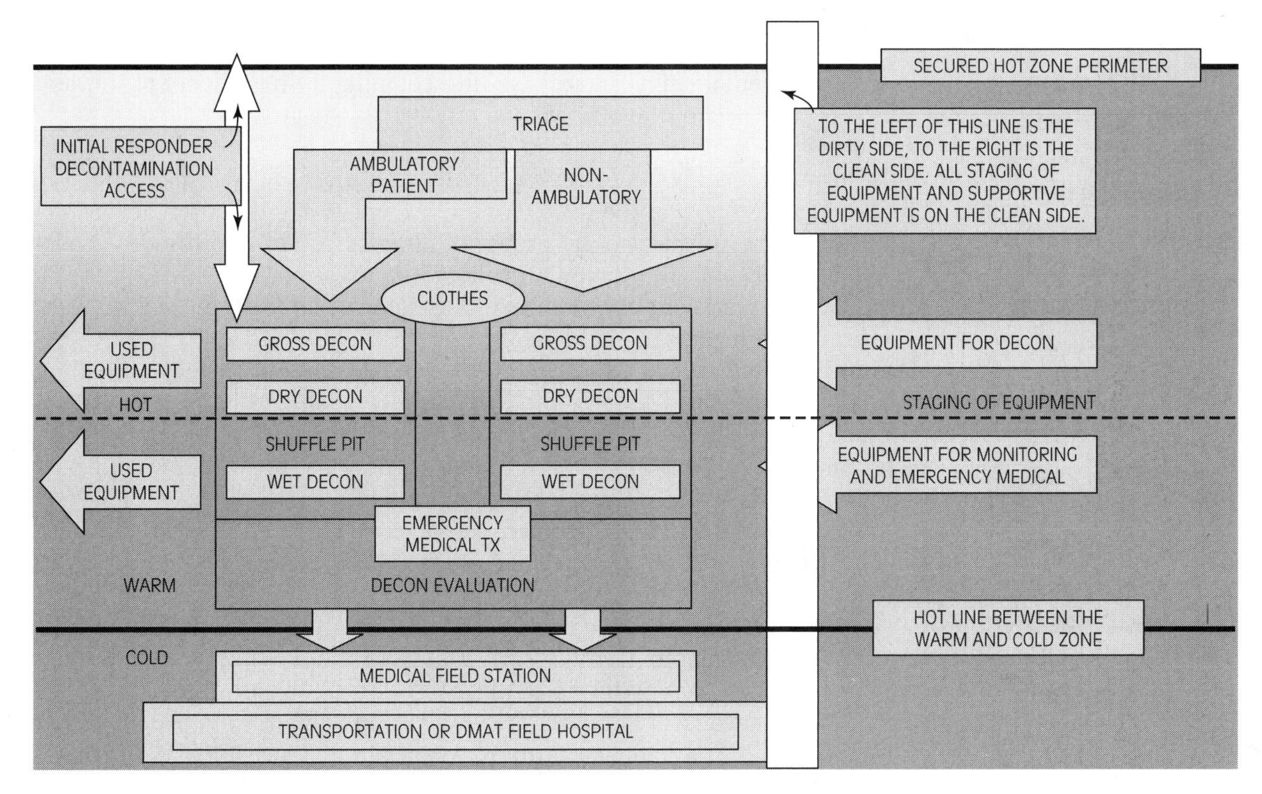

Figure 5-6 *A decontamination corridor.*

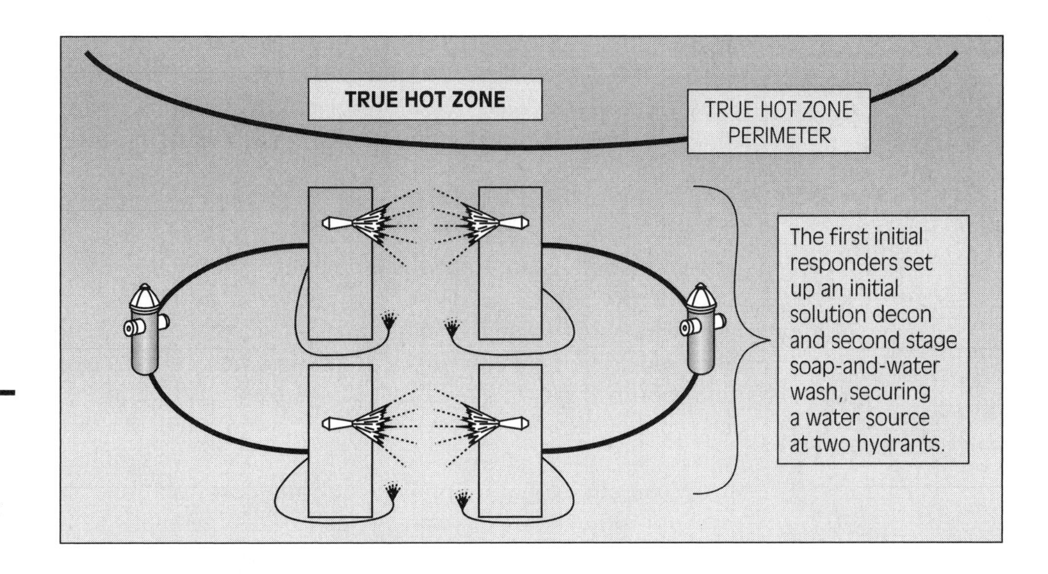

Figure 5-7 *A close-up of a decontamination corridor near a hot zone.*

Levels of Impact

■ Note
At this point, efforts should be directed to controlling fires, protecting lives on the ground, and preserving evidence.

■ Note
The likelihood of a post-crash fire is great. There is a good chance of survivors, but they are likely to have sustained moderate to severe trauma.

emergency escape route
a safe escape path for passengers and crew of an aircraft crash landing to quickly escape with minimal chances of burns or sustaining secondary injuries; for firefighters or other emergency responders, a designated escape route to exit a "danger zone" in the event it becomes too dangerous to continue their involvement in the operation

As you size up an emergency scene, take into consideration the **level of impact**. This refers to the speed of the aircraft when it strikes the ground.

High-impact crashes are characterized by extensive breakup and disintegration of the airplane. Most of the wreckage is in small pieces. The impact site may contain a crater. There may be secondary fires and a combination of various gases in the smoke (discussed in previous chapters). The likelihood of aircrew/passenger survival is extremely unlikely in a high-impact crash.

At this point, efforts should be directed to controlling fires, protecting lives on the ground, and preserving evidence.

Medium-impact crashes are characterized by the aircraft fuselage breaking into several large pieces. The likelihood of a post-crash fire is great. There is a good chance of survivors, but they are likely to have sustained moderate to severe trauma. In medium-impact accidents, accessing trapped crew and passengers often entails forcible entry into the aircraft.

Cut into an aircraft at clearly marked cut-in areas. If these are not visible, cut between or just above passenger windows, because these areas are least likely to have cables or hydraulic lines running through them. The dangers of cutting through a hydraulic line have been discussed previously, but cutting through an aircraft control cable is potentially equally dangerous. If the cable is cut while it is under stress, it may snap with violent force.

Ideally, use a power saw or an *air chisel* to cut into an airplane. An air chisel is handy for cutting through the aluminum exterior (see **Figure 5-8**). (When using a power saw to cut into an aircraft, is it best to use carbide blades.) Make your cuts just beneath the surface of an aircraft. A charged hoseline should be in place to suppress any heat or sparks generated by this operation. If a power saw or an air chisel is not available, make a cut-in by placing a large flathead screwdriver against the aluminum exterior at a 45-degree angle. With a hammer, strike the edge of the screwdriver about a half-inch away from where its tip contacts the aluminum. Apply repeated blows from the hammer as you move the screwdriver along the line of your intended cutting path. This low-tech technique is surprisingly effective.

Once you have gained access to the aircraft's interior, searches for survivors and the deceased must be conducted cautiously and meticulously. Survivors likely must be disentangled from cables, electrical wiring, and assorted internal and exterior wreckage. Rescue tools, such as *come-alongs* and *porta powers,* maybe have to be used, because aircraft seat mountings, floors, and so on tend to bend and give way easily. In some crashes, small children have been found thrown into overhead passenger cabin storage bins, and people have been found in the cargo hold below the main cabin floor, which has partially collapsed. The most common injuries to medium-impact crash survivors include burns, smoke inhalation, head and spinal trauma, lacerations, and fractures.

Low-impact crashes can occur either on or off of an airport runway; off-airport crash landings may take place on a county road or in an open field. These

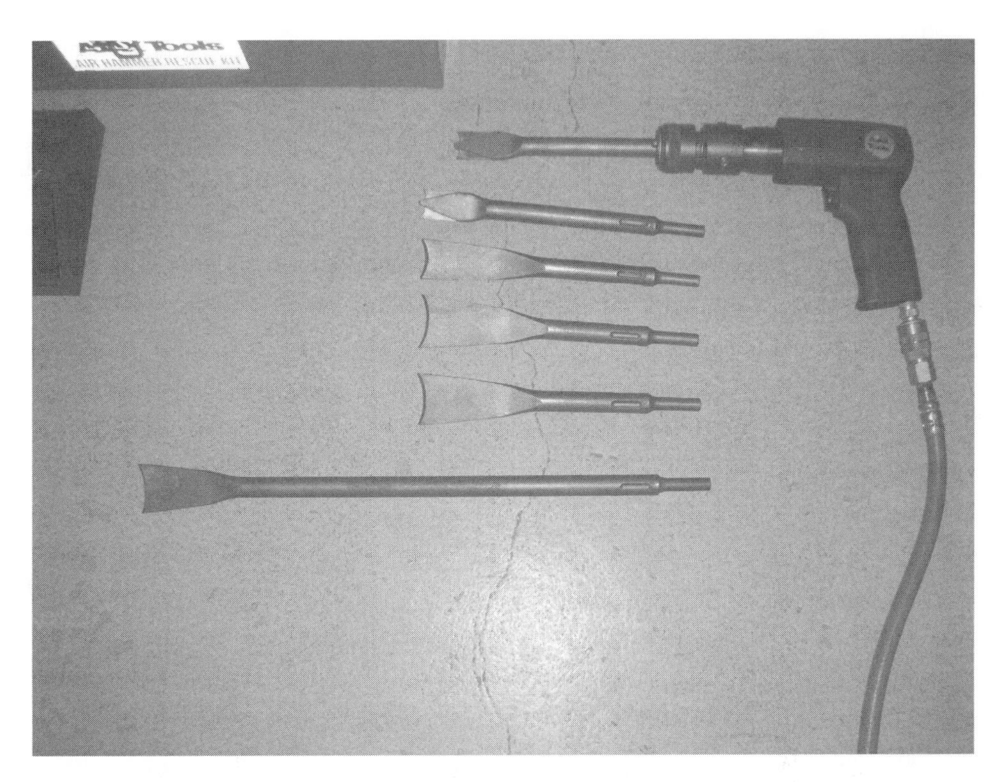

Figure 5-8 *An air chisel.*

crashes are characterized by a basically intact fuselage and many survivors. You probably will encounter injured people and fires on the ground as well as inside the aircraft.

■ Note **Quick response, rapid fire suppression, and aggressive rescue actions increase the chances of survival for people trapped inside an aircraft.**

Quick response, rapid fire suppression, and aggressive rescue actions increase the chances of survival for people trapped inside an aircraft. Anticipate the need to provide care for injured people inside the airplane as well as those who have been thrown clear of the airplane.

CASE STUDY

In February 2005, a business-corporate jet experienced a low-impact crash during attempted takeoff from an airport in Teterboro, N.J. The airplane landed, crossed six lanes of a busy highway, and then crashed into the side of a warehouse and caught fire. All passengers and crew on board the stricken jet survived. A coordinated operation among airport and non-airport firefighters, law enforcement, and other disaster response agencies provided an example of an effective unified command.

tactics
the "doing" of all the necessary operational activities needed to accomplish the goals set by the IC; must follow the goals and objectives set forth by the strategic plan

level of impact
the speed of the aircraft when an aircraft struck the ground

high-impact crash
characterized by extensive break-up and disintegration of the airplane

medium-impact crash
characterized by the aircraft fuselage breaking apart into several large pieces

low-impact crash
can be either on or off an airport runway such as a county road or open field; fuselage is basically intact

Surfaces and Terrain

Rescue approach and operations on a hard surface are the best-case scenario in an aircraft incident. Nonetheless, even these optimal conditions pose dangers. For example, if a crash site occurs on concrete or asphalt, spilled fuel and fire can weaken these materials, rendering them dangerous for your firefighters and vehicles to operate on. Intense fire may cause concrete to spall (break apart) when water or foam is applied and project chunks of concrete at responders and equipment with great force.

Terrain is another major factor in dealing with an aircraft incident. As previously mentioned, crashes can occur anywhere. Anticipate that trees, power lines, or utility poles may have been struck down, impeding your accessibility to the site and affecting your tactical operations. Fires burning directly beneath power lines may weaken them and cause them to break and fall to the ground. Streams and waterways will hinder your access to a site, although they may also prove to be a valuable source of firefighting water. Hills may cause apparatus to lean at dangerous angles during approach and stationary operations, depending on the angle of the hill and the off-road capabilities of your emergency rescue vehicles.

If an incident occurs during daylight, you have a much better chance to read the topography than if it happens at night. No matter how well you know an area, it appears entirely different in darkness or during inclement weather, such as heavy rains or snowstorms. In one instance, rescuers were hindered by a heavy rainstorm punctuated by lightning, severe winds—and snakes that were retreating from lowland areas adjacent to a river.

FIRE-EXTINGUISHING AGENTS

For every type of fire, there is an appropriate type (or best choice) of fire-extinguishing agent to use on it for maximum effectiveness. Know what agents you have available and the proper application technique of those agents for the class of fire you are dealing with and the variable conditions associated with that fire.

For example, **Class A foams** are penetrating agents used on matted vegetation in wildland fires, while **Class B foams** are used to extinguish fuel fires. Live burn tests on carbon-fiber composites have shown that Class B **aqueous film forming foam (AFFF)** is the best agent for extinguishing burning composite materials. When fighting Class B fires involving polar solvent fuels containing any form of alcohol, alcohol-resistant foam is the agent of choice.

Dry chemical agent is versatile in many firefighting applications. When combating Class C (electrical) fires, dry chemical agents may be effective, and may leave residue that may damage electrical equipment. Gaseous fire-extinguishing agents, however, including carbon dioxide and agents in the Halon family, are very effective for fires involving electrical equipment and do not leave residue.

Danger Using water/foam on energized electrical equipment may be extremely dangerous.

Using water/foam on energized electrical equipment may be extremely dangerous. For fighting Class D (metal) fires, a dry powder agent is best.

During an aircraft fire, attempt to prevent reignition by shutting off power to the aircraft's electric equipment by disconnecting the battery. If you can use aircraft emergency shutdown T handles, remember that these may not function as designed without electrical power. *Never* use Halon on burning magnesium or carbon-fiber composite materials, because these agents displace oxygen. When using these agents, wearing a self-contained breathing apparatus is a must.

Warning Never use Halon on burning magnesium or carbon-fiber composite materials, because these agents displace oxygen. When using these agents, wearing a self-contained breathing apparatus is a must.

Water Fog/Spray

Water fog/spray is commonly used during a structural fire emergency or on fires involving nonpetroleum products. Water is normally plentiful and the least expensive agent with which to combat a fire. Water fog or spray may not be effective during an aircraft incident, however. If fuel is the source of an aircraft fire, water will cool the area, but may also spread the fuel to uninvolved portions of the aircraft, thus actually increasing the area involved in the fire. If water is your only option, it may be applied using interlocking high-pressure stream patterns. Most airport firefighters, as well as non-airport firefighters, have been trained in this method of extinguishing a large Class B fire. This is the most dangerous method of extinguishment, however, and firefighters should receive recurrent training in it.

Water also is a poor choice for extinguishing burning magnesium. This metal burns with an intense heat and bright, and once this burning metal comes in contact with water, it breaks down into its two basic elements, hydrogen and oxygen. The hydrogen is extremely flammable and will intensify the fire, as will the freshly liberated oxygen. When burning magnesium comes in contact with water, it commonly blows apart, scattering molten fragments with explosive force.

A reliable and abundant supply of water is needed at aircraft incident scenes, however, for decontamination and the mop-up phase of the fire-suppression operation.

Class B Firefighting Foams

The most common Class B firefighting foam used in aircraft firefighting is aqueous film forming foam (AFFF). AFFF creates a protective blanket that allows fire to be extinguished. And, once a foam blanket has been established, it forms a vapor barrier minimizing the chances of a reignition. In contrast to older versions of foam, AFFF can be self-sealing, meaning that if the foam blanket is disturbed, it is more efficient than previous versions at re-closing and reforming a complete blanket.

Note It is crucial that foam blankets be monitored and maintained, because blankets can be compromised by many factors.

It is crucial that foam blankets be monitored and maintained, because blankets can be compromised by many factors. For example, the water in the foam gradually drains, depending on the amount of foam applied and the relative

class A foam
firefighting foam that is mixed with water, allowing the extinguishing agent to penetrate deep-seated fires involving matted grasses or other dry vegetation, bales of cotton, cardboard, or other Class A combustibles requiring a penetrating fire-extinguishing agent; also used to temporarily coat exposed flammable vegetation and structures, affording temporary protection from fire

Class B foam
used to control and extinguish fuel fires (Class B fires), such as those caused by gasoline or other liquid hydrocarbon fuels

aqueous film forming foam (AFFF)
fire-extinguishing agent that contains fluorocarbon surfactants and spreads a protective blanket of foam that extinguishes liquid hydrocarbon fuel fires by forming a self-sealing barrier between the fire and fire-sustaining oxygen

humidity. Water from rain or firefighting mop-up/salvage and overhaul operations may dilute the concentration of foam, thereby degrading the blanket. A blanket also can be disrupted by objects being dragged across it, foot traffic, or vehicles being driven through it. In addition, wind may affect a foam blanket by blowing it away or accelerating evaporation of the water in the foam.

If a blanket weakens, regardless of the cause, the flammable liquids the blanket is covering can emit potentially ignitable vapors or smoldering materials it is protecting (such as very hot magnesium) can violently rekindle. If you observe gaps in the foam blanket, or notice it begin to dry and blow away, reapply foam to the area, keeping the blanket intact. As a rule of thumb, many agencies reapply the foam blanket every 20 minutes to ensure consistent coverage and efficacy.

AFFF can be delivered in different percentages, ranging from a 1 percent- to an 8 percent-ratio mixture of foam product to water. Know the appropriate application percentages and techniques for different fire types.

AFFF can be applied using a **raindrop effect**, in which the stream is elevated allowing the foam to cascade downward onto the fire. The resulting foam naturally forms a blanket over spilled fuel fires, extinguishing the flames. Another means of application is rolling the foam stream like a wave across the fuel fire and banking it off a surface, such as a piece of the fuselage, nearby rock or similar surface.

A good example of the "raindrop effect" technique is shown in **Figure 5-9,** using the principles of stream elevation, motion, and wind compensation. In this training exercise, the vehicles changed stream patterns as they got closer to the flames. In this picture, the ARFF crews are keeping the aircraft cooled as they simultaneously sweep their turret streams to "rain" AFFF on the fire.

Dry Chemical Firefighting Agents

Dry chemical agents are very effective alone or when used in combination with interlocking water fog nozzle patterns, or with AFFF. The preferred method of attacking a running fuel fire with a dual-agent response, such as using both AFFF and dry chemical. Dry chemical agents come in a variety of types for specific and general applications, in other words, Purple K (potassium bicarbonate) is used for class B and C fires only while monoammonium phosphate (a typical ABC dry chemical) is used for any Class A, B, or C fires. Know what you have available to you before you are placed in a situation to use them to save a life and extinguish a fire.

Figure 5-10 illustrates some fundamental, generic firefighting procedures for several kinds of aircraft fires that do not require a major incident command structure, and that use agents such as carbon dioxide and Halon. Although dry chemical agents work on these fires, they ruin equipment and make a mess. *Never* use these agents on wheel fires. The illustration shows the application of dry powder agent for wheel fires (e.g., a Class D agent for a Class D fire).

Figure 5-9 *An ARFF vehicle demonstrates proper technique for applying AFFF using the "raindrop effect."*

Dry Powder Firefighting Agents

Dry powder agents normally are used on Class D fires, fires involving metals. It takes a lot of heat to ignite metal, but metals do burn when they reach a certain temperature. Because such intense heat is required to make metal burn, metal fires require a lot of dry powder to extinguish. Ensure that you are using an NFPA-approved self-contained breathing apparatus whenever operating around a dry-powder extinguishing agent.

• Caution
Thus, when sand is applied to burning magnesium, its moisture quickly expands into steam, which blasts the sand in all directions and can cause injuries.

It is commonly believed that sand can be safely applied to burning metals, such as magnesium, instead of a dry powder agent. All sand, however, with the exception of truly dry sand, contains moisture. Thus, when sand is applied to burning magnesium, its moisture quickly expands into steam, which blasts the sand in all directions and can cause injuries. Only sand that is *truly dry,* which is stored in specially marked bags, is safe to use on metal fires. Even bagged dry sand, however, still is potentially dangerous to rescuers, because it contains silica, a known carcinogen, and because the fine granules of sand irritate the respiratory tract. Thus, whenever using sand, you must wear some form of respiratory protection.

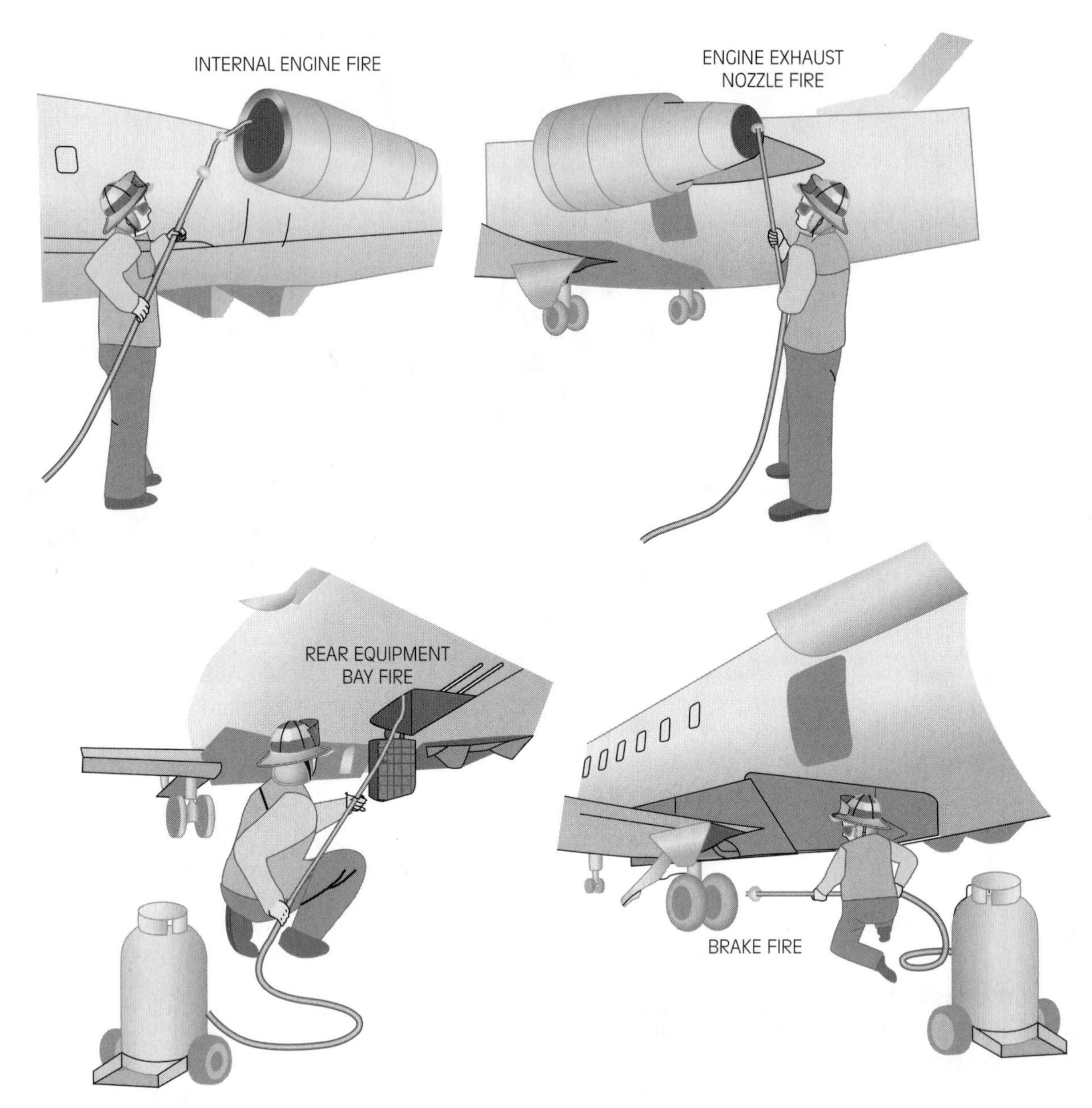

Figure 5-10 *Applying CO_2, Halon, dry chemical, or dry powder agents with an extinguisher extension wand provides greater reach.*

raindrop effect
the most efficient aqueous film forming foam (AFFF) firefighting technique; the stream is elevated allowing the foam to cascade downward onto the fire

dry chemical
a fire-extinguishing agent used to extinguish Class A, B, or C fires and composed of ammonium phosphate (ABC), potassium bicarbonate (brand name, Purple K), or similar material

dry powder
fire-extinguishing agent used for Class D (metal) fires and usually made of powdered graphite or similar materials; brand names include Metyl-X™

Another substitute for a commercial dry-powder extinguishing agent is dry Portland cement, which does not react with burning metals. Cover the burning magnesium with the dry cement, permitting it to form a cake or pile around the metal.

ASSISTING AIRPORT FIRE DEPARTMENTS

Your agency may be called on to help an airport fire department for a wide array of reasons. You may be tasked with actual firefighting, rescue, or resupply operations. Regardless of what your role is, you can better serve in it if you have a general understanding of the methods used by the department you are assisting.

Firefighting

Normally, fighting aircraft fires at an airport is a task for firefighters who have special training in airport rescue and firefighting operations and are equipped with the appropriate personal protective clothing. If you are near a small airport, however, the available numbers of these trained professionals may be limited; some regions may have no ARFF-certified firefighters available. Therefore, your agency may be called in to assist with aircraft fire-suppression operations at a small airport.

If your agency is equipped with only structural firefighting clothing, however, tell the incident commander this before engaging in a dangerous operation. Many tasks that can be safely accomplished wearing structural firefighting protective clothing are instrumental to the successful outcome of an accident.

In most situations where airport fire crews are available, these crews manage the aircraft fire-suppression operation, while local fire departments handle structural or vegetation fires.

Rescue

■ Note
Safety is your top priority. Scene assessment is an ongoing process, and any changes in hazards or tactical priorities must be relayed to everyone involved in the operation.

Aircraft accident victims who are capable of movement will escape on their own, often long before emergency responders arrive at the scene. These people should be taken to the medical triage/treatment area as soon as possible. Try to gather as much information as possible, from survivors, especially from surviving aircrew members. Examples of valuable information include the number of people onboard the aircraft, known locations of people trapped inside the airplane, and any known cargo (and weapons systems on combat aircraft).

If you are asked to help free personnel still trapped inside the wreckage, use the same priorities as you would use during a structural fire or an automobile accident.

Prioritize rescue and medical care according to your local protocols. Safety is your top priority. Scene assessment is an ongoing process, and any changes in hazards or tactical priorities must be relayed to everyone involved in the operation.

Resupply

Resupply operations are paramount to successful rapid fire extinguishment during a large-scale aviation accident and post-crash fire. Fire-suppression efforts become ineffective if firefighting apparatus have run out of extinguishing agent and have to disconnect and fall back for reservicing. Ideally, apparatus that are directly fighting an aircraft fire should be resupplied while they are in place, to provide a continuous water supply. You can resupply airport fire apparatus directly from your own apparatus, perform a water-shuttle operation, or establish a continual water supply from a nearby hydrant system. If none of these methods is possible or sufficient, you may have to perform relay operations from the water source to the frontline apparatus, as shown in **Figure 5-11.** This picture shows a rural fire department training to control an aircraft fire using a tanker shuttle resupply technique. Not visible is an additional water-tender that is approaching to connect a supply line to a clapper valve, ensuring a continuous water supply after the first water-tender runs out of water. This also is referred to as a nursing operation.

Figure 5-11 *A simulated fire attack with resupply in progress using structural/brush apparatus and water tenders.*

MASS CASUALTY ISSUES

Aircraft accidents often generate mass casualty incidents. Depending of the level of service requested of your agency by an airport fire department, you might be required to set up a triage area, triage and litter (move) patients, or perform some level of medical treatment. You may also be asked to gather and contain human remains. Regardless of which of these tasks you are assigned to, you must protect yourself physically and medically from injury and illness.

universal precautions
work practices that comply with OSHA regulations and include wearing body substance isolation equipment and clothing, engineering controls (built-in protection), wearing personal protective equipment (PPE), decontamination, and properly disposing contaminated materials

body substance isolation (BSI)
wearing personal protective equipment (PPE), such as gloves, goggles, gowns, and masks, while rendering first aid or in any other way risking exposure to any human body fluids

occupational exposure
unprotected contact with potentially infectious materials that may occur during the performance of an employee's duties

Protecting Yourself

When dealing with a mass casualty incident, taking **universal precautions** is required by federal, state, and local governmental regulations. Per OSHA regulations, universal precautions are *mandatory* whenever there is *anticipated exposure* to bodily fluids, because ordinary clothing, such as uniforms or work pants and shirts or blouses, are not designed to be or rated as being protective against a disease-carrying body fluids.

Universal precautions include wearing appropriate personal protective equipment (PPE), such as gloves, goggles, gowns, and masks, to achieve **body substance isolation** while rendering first aid or otherwise risking exposure to human body fluids. The terms *universal precautions* and *body substance isolation* refer to avoiding contact with patients' bodily fluids by wearing nonporous PPE. When taking universal precautions at a crash scene, be especially careful around sharp medical instruments, jagged wreckage, and broken glass, all of which may bear human bodily fluids. (Hypodermic needles, particularly, should be handled carefully and disposed of properly in a *sharps container.*)

Personal protective equipment for universal precautions is clothing or equipment worn to prevent contact with human-, animal-, or insect-transmitted pathogens, or germs. *Pathogens* fall into two categories: blood-borne (carried in body fluids) and airborne. Standard universal precautions protect rescuers from both types of pathogens.

Occupational exposure occurs when a person is exposed to a pathogen source (such as potentially infectious materials) while performing his or her job duties. If the pathogen enters the person's body, it may spread disease. Universal precautions should be practiced in any environment where workers are exposed to bodily fluids so as to prevent disease contraction. Diseases that exposed persons may develop include, but are not limited to, HIV, hepatitis, and meningitis, among others.

To avoid exposure to these and other diseases, it is safest to not directly come in contact with anything that has been inside the human body. Blood, saliva, semen, and vaginal fluids all can transmit disease, but other bodily fluids can as well, including cerebrospinal, pericardial, synovial (joint), and amniotic fluids, In addition, substances such as urine, feces, vomit, perspiration, and sputum may carry pathogens. These materials may be spattered on smooth and jagged aircraft wreckage, cabin interiors, floors, walls, escape slides, and lavatories.

■ **Note**
Not all Tyvek® suits are nonporous. Ensure that any clothing used as PPE is approved for the prevention of exposure to bodily fluids and pathogens.

Whenever emergency responders may be exposed to human body fluids, they should wear nonabsorbent, nonpermeable clothing, as well as the appropriate footwear and eye protection. Not all Tyvek® suits are nonporous. Ensure that any clothing used as PPE is approved for the prevention of exposure to bodily fluids and pathogens. Over-shoe booties or rubber boots should be worn, depending on the site situation. In addition to wearing the proper equipment, you must remember to change gloves after every patient to prevent cross-contamination, which occurs when germs from one injured survivor are accidentally transmitted to another survivor by a responder.

Universal precautions also include properly decontaminating both personnel and PPE. Implementing all appropriate universal precautions will keep you safe from many blood-borne, vector-borne, and airborne diseases. (Additional information regarding universal precautions and blood-borne pathogens are found in 29 CFR Blood-borne Pathogens–1910.1030.)

■ **Note**
It is imperative that you understand and train with the triage system employed in your community *before* an incident occurs. During an incident is not the time to learn or practice these procedures.

What to Expect

At the scene of an aircraft crash, expect the unexpected. There are no textbook aircraft accidents. Many will be similar in nature, but there will also be differing tactical considerations due to the individual characteristics of each specific incident. A military airplane crashing onto a bank parking lot poses a different set of challenges than a commuter plane that crash lands in a farm field. Be ready for anything: Murphy's Law applies. If you start the incident with little lateral wiggle room on your tactics, you may inadvertently commit the firefighting/rescue assets to a losing battle. Weather and wind conditions affect the scene dramatically just to name one variable. Expect the worst and hope for the best. As you implement Plan A, have Plan B ready. Always have a secondary plan in case something changes the nature of the incident you are attempting to manage.

Triage

triage
the process of sorting patients for priority treatment based on factors such as severity of injury, likelihood of survival, and available resources

Triage is the task of sorting patients according to the severity of injury they have sustained. Systems of designation include ranging systems (from "good to grave" or "1 to 5") and color codes; use whichever system is common to your area. It is imperative that you understand and train with the triage system employed in your community *before* an incident occurs. During an incident is not the time to learn or practice these procedures.

Airports may have passenger care teams comprised of volunteers from the airlines, local mental health teams, and the American Red Cross. These people are designated to deploy at a moment's notice to aviation disasters involving passenger-carrying aircraft. They work in conjunction with other volunteer organizations arranging assistance for survivors, such as meals, clothing, lodging, communication, and transport of family members to the accident site as needed. They also address various other needs that may arise affecting either surviving passengers or their families.

In addition to dealing with injured survivors, you may have to deal with the remains of the deceased. These are gruesome tasks, but it is important to remember to treat any human remains with dignity.

It may be necessary to move a body in order to reach a trapped survivor. Or, bodies or portions of bodies may be strewn around the debris field and need to be gathered. Human remains may be in trees, floating in water, or being eaten by animals.

If tasked with retrieving human remains, wear latex, nitrile, or vinyl gloves beneath thick leather work gloves. Cover the remains as soon as possible with tarps. Make sure the locations of any human remains are marked with small flags or other markers as appropriate.

HELICOPTER OPERATIONS

Seriously and critically injured crash survivors may require immediate transport to the closest medical facility via aeromedical evacuation, or *medevac*. Most **aeromedical operations** involve helicopters, which can get into and out of crash sites more easily than other types of aircraft. Emergency helicopter operations require orchestrated teamwork among various personnel and agencies, including medevac aircrews, fire and police, ground-based EMS personnel, or military agencies.

aeromedical (medevac) operations
emergency aircraft (usually helicopter) operations that require teamwork among different agencies, including medevac aircrews, fire, police, ground-based EMS personnel, or military agencies

This is partly because helicopter crews need assistance from ground personnel to select a safe landing zone (LZ) close to the patient needing transport. Ground operations include identifying and communicating to pilots possible hazards and keeping bystanders away from the landing zone. This is important, because the presence of bystanders can compromise the safety of aircrew, emergency responders, and the onlookers themselves.

Planning for Aeromedical Evacuation (Medevac)

An emergency helicopter landing zone is basically a temporary heliport: In addition to medevac, it may be used as a base for airborne firefighting operations, search and rescue missions, or emergency crew and equipment transportation. Your agency's personnel may be needed to select a safe helicopter landing/takeoff site, supply emergency standby crews, or arrange for the refueling of the helicopter and emergency rescue vehicles.

In 1998, the FAA Office of Communications, Navigation, and Surveillance Systems published an advisory circular, AC No. 00-59: *Integrating Helicopter and Tilt Rotor Assets into Disaster Relief Planning,* to assist state and local emergency planners in safely executing aeromedical evacuation. Review this publication, as it will provide you with general guidance for integrating helicopters and tilt-**rotor** aircraft into your disaster relief Standard Operation Guidelines (SOG). The FAA recognizes, however, that every state and municipality has

rotor
rotating airfoil assembly (propeller) of helicopters

distinctive and unique requirements that may warrant modifications to the advice found in the circular.

In general, however, your SOG should include plans for setting up communications among ground and air crews, such as establishing distinct emergency frequencies for air operations, ground operations, and medical/rescue information exchange. Use separate channels for these if possible, as this facilitates efficient management of aircraft operations and the medical/rescue network. Segregating the communications net helps prevent the potential conflicts created by sharing radio frequencies.

If a large-scale disaster, a communications equipment malfunction, an act of terror, or a civil disturbance disrupts regular telephone service, try using satellite communications or microwave cellular phones. Also, know the telephone number (including cell phone number) of the aeromedical dispatch base operations or of a nearby airport **fixed-base operator** that may be able to maintain phone communication to the aircraft.

fixed-base operator (FBO) entity at an airport, which renders maintenance, storage, and servicing of aircraft; may also rent aircraft to licensed pilots

Helicopter pilots communicate with landing zone personnel to obtain critical information to ensure safe landing and departure. Thus, any personnel involved with medevac operations must be familiar with radio communications techniques, including the aviation **phonetic alphabet**, and with the proper landing hand signals (discussed later; also see Figure 5-14). When communicating with aircraft, ground crew should use vehicle radios, instead of portables, to obtain maximum signal range and clarity.

phonetic alphabet spoken-word alphabet designed to eliminate confusion for listeners when letters sound alike

Establishing a Helicopter Landing Zone

A safe **landing zone (LZ)** for helicopters should be a *minimum* of 100 feet by 100 feet (although some regions permit zones of 75 feet by 75 feet). The LZ also must be located at least 500 feet from the crash site. This is to prevent destruction of the firefighting foam blanket and the scattering of lightweight equipment and accident evidence. Coordinate and preplan with your local aeromedical agencies, and include this requirement in your plan (see **Figure 5-12**). This requirement also applies to **tilt-rotor aircraft**; however, you need greater dimensions for tilt rotors or larger helicopters. Never select a frozen body of water for a landing zone. Although very thick ice can develop in some geographic regions, hot auxiliary power exhaust may thaw and fracture even solid ice. The landing surface should have a less than a 10-degree slope and be firm, so that mobile equipment, rescuers, or the aircraft itself don't become bogged down. Clearly, deep snowdrifts and areas of deep mud are inappropriate for landing zones.

landing zone (LZ) an area designated for the transfer of medical patients or evacuees from a ground area to board an aircraft, usually a helicopter

tilt-rotor aircraft vehicles that can take off, land, and hover similarly to a helicopter, and then can change wing angle and fly like a turboprop airplane; also called "power-lift vehicles"

Ground crew should select an LZ containing as few obstacles and other hazards as possible. Often, hazards are not seen from the air as easily as they are from the ground, so the ground crew also should advise pilots of the location of all obstacles. Do not establish an LZ close to radio towers or antennas, trees, power poles, telephone wires, or other obstacles that would then be in the helicopter's approach and departure path. The LZ also should be clear of any low objects that

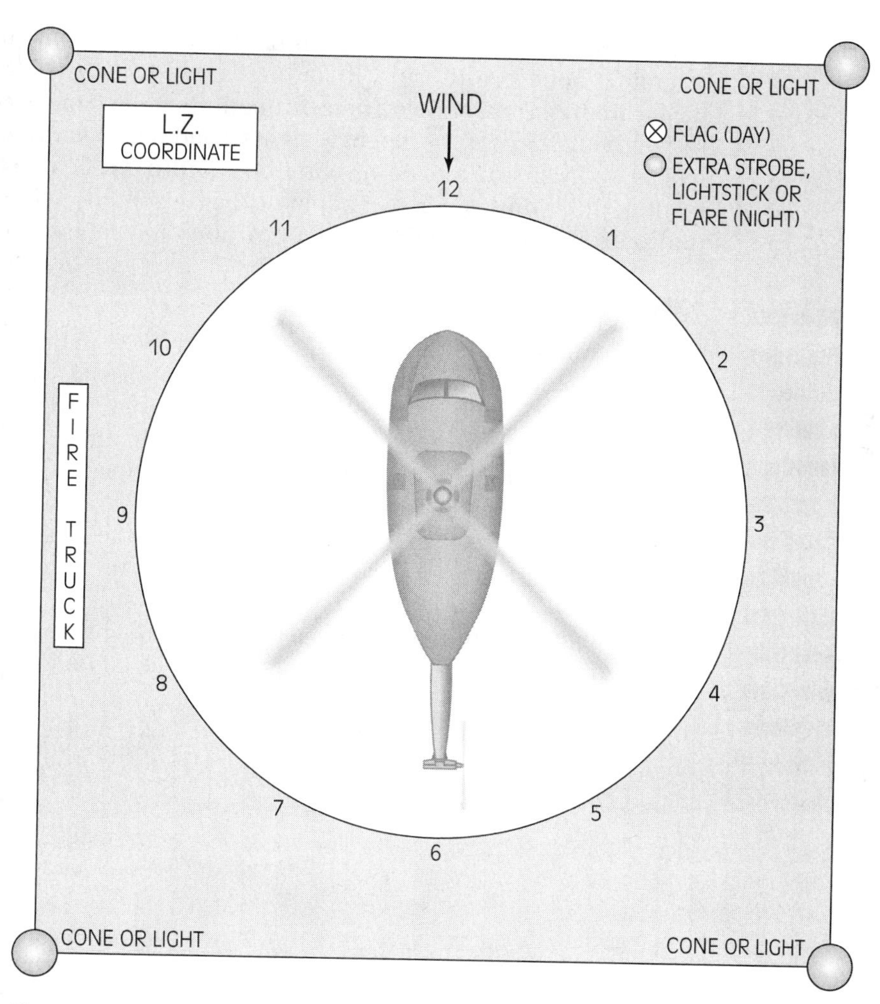

Figure 5-12 *A helicopter landing zone.*

> **Warning**
> Never select a frozen body of water for a landing zone. Although very thick ice can develop in some geographic regions, hot auxiliary power exhaust may thaw and fracture even solid ice.

may get caught on a helicopter skid, such as fences and livestock corrals. Approach and departure paths should not pass over command posts, treatment areas, or operationally congested ground areas where rotor wash and/or noise may interfere with communications and operations. Last, helicopter exhaust may deposit on nearby automobiles and other objects, so, for this and other safety reasons, advise nonessential people (i.e., the public) to park their automobiles away from the LZ.

Another concern when determining the location of an LZ is that rotor wash—which can exceed 110 mph—may damage buildings made fragile from a natural disaster or terrorist attack. Rotor wash may also be hazardous to crashed aircraft, tents and other temporary structures, damaged overpasses, or wrecked automobiles.

To provide optimal safety at an LZ, the incident command should designate a landing zone coordinator. If staffing is adequate, the LZ coordinator should have three assistants: Two should function as left- and right-perimeter guards, respectively, at the midpoint of their assigned perimeter lines after the helicopter has landed. The third remains with the coordinator to assist with moving equipment or other duties. Some aeromedical agencies allow helicopter pilots to lock the controls at idle speed and then position themselves as a tail rotor guard.

■ Note

Ensure that fire crews are fully clothed in turnout gear before a helicopter lands and during its departure. The apparatus fire pump should be engaged, and charged fire hoses must be readily available. Class B firefighting foam should be on hand if possible.

Emergency Standby Crews

Many communities require fire apparatus and crews to be present at *all* emergency helicopter landings, to ensure safety. If you are tasked with standby for an emergency helicopter landing, position your fire engine on the side of the LZ perimeter away from the approach and departure paths of the helicopter. Then, soak the landing area to contain dust, dry hay, and so on (green, grassy landing zones do not require this). Firefighters should position their fire apparatus between them and the helicopter as it lands or takes off, to protect the rescue crew if the helicopter crashes and generates an explosion, fireball, or flying debris. Ensure that fire crews are fully clothed in turnout gear before a helicopter lands and during its departure. The apparatus fire pump should be engaged, and charged fire hoses must be readily available. Class B firefighting foam should be on hand if possible.

When firefighters frequently participate in uneventful aeromedical standby operations, they may become lax in these safety procedures. Do not allow your personnel to let their guard down. **Figure 5-13** illustrates the danger areas associated with helicopter operations.

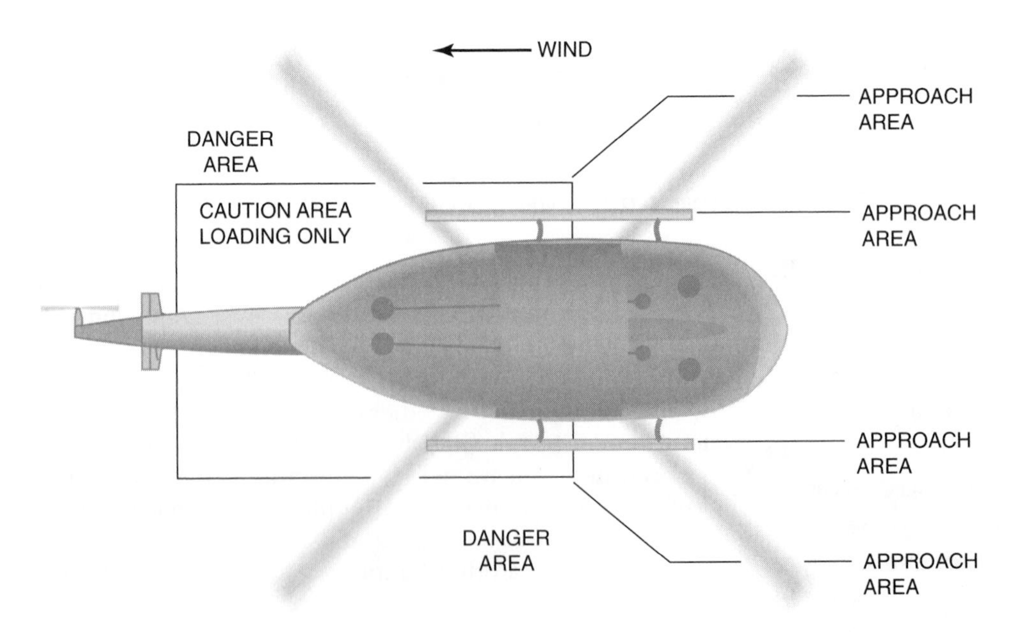

Figure 5-13 *Common helicopter danger zones.*

Helicopter Landing Zone Coordinator's Checklist

Before Landing

- ❑ Select a flat and firm surface for the LZ.
- ❑ Conduct a briefing to your LZ assistants.
- ❑ Ensure there are no tall obstacles that may endanger the helicopter landing or departure.
- ❑ Advise ground and air crews the location of the LZ.
- ❑ Ensure LZ perimeter is kept clear of non-essential personnel using law enforcement, firefighters, or **augmentees**.
- ❑ Mark the LZ according to *day* or *night* operations.
- ❑ Ensure fire emergency standby vehicle is in place, crews in PPE.
- ❑ Ensure no hats are worn in the LZ. Chinstraps on helmets must be tight enough to protect loss from rotor wash.
- ❑ Use eye protection and hearing protection.
- ❑ Ensure the LZ is free of any loose materials that might be blown into the rotors or engines. Wet down any loose sand, dirt, bit to prevent dust clouds and other flying debris. Designate a ground unit to communicate with the aircraft.
- ❑ When establishing communication with the helicopter, give a brief update of the situation such as changes in the landing zone, describe all potential hazards, and update patient information if it has changed significantly.
- ❑ Advise helicopter pilots when their aircraft are in sight.
- ❑ Direct the pilot to the scene, using true north compass bearings, and geographical landmarks.

augmentee
person who volunteers, or is recruited, to assist in an emergency operation; can be utilized for non-hazardous duties, such as assisting with crowd control, obtaining ice, assisting in cutting wood for cribbing, and so on

After Landing

- ❑ Keep all nonessential emergency personnel and the general public at least 100 feet from the landing area.
- ❑ All equipment, stretcher sheets, IV poles, pillows, and so on must be secured against rotor wash.
- ❑ Always approach the helicopter from the front, downhill, within full view of the pilot, never from the rear except helicopters that are designed to load or remove patients from the rear. Make eye contact with the pilot. If you don't see the pilot, the pilot doesn't see you.
- ❑ Do not permit personnel to run within the LZ.
- ❑ Do not approach the helicopter until signaled by the aircrew.
- ❑ Do not take the patient to the helicopter (unless the aeromedical crew directs you to do so): Allow the aeromedical crew to come to you.
- ❑ Carry equipment horizontally, below your waist level—*never* carry it upright or over your shoulder. Ensure that IV poles are clear of rotors.

(*continued*)

Helicopter Landing Zone Coordinator's Checklist (*Continued*)

- ❑ The flight crew must direct the loading and unloading of the patient and equipment.
- ❑ Doors or hatches on the helicopter should be secured only by the aircrew because they can be easily damaged.

Liftoff and Departure

- ❑ The landing zone coordinator notifies the pilot when the landing zone is clear of all ground personnel.
- ❑ Continue protective measures against flying debris.
- ❑ If at all possible, try to maintain a secure landing zone with all personnel and emergency equipment for at least 5 minutes after the helicopter departs. If an in-flight emergency develops, this allows the pilot to return safely to a secure landing zone.
- ❑ Release personnel and equipment assigned to the LZ on directions of the incident commander.

Safety Procedures Whenever you must approach a helicopter, practice these safety precautions:

- Always walk, never run, around a helicopter.
- Approach and assist with patient loading *only* when requested to do so by the aircrew.
- Always approach or leave a helicopter from downhill (especially if the landing zone is uneven), in full view of the pilot, between the 10 and 2 o'clock positions of the front of the aircraft. This is to avoid engine or APU exhaust and the tail rotor, which usually is located at head level and is almost invisible. Poor lighting conditions contribute to tail rotor danger.
- When approaching the aircraft, always make sure the pilot sees you and waves you forward.
- Remain in a low, crouched, position at all times. This is to avoid injury or death from a phenomenon called *flap* or *sail*. This phenomenon can occur while the rotor RPM speed up or slow down. A sudden wind gust can cause the rotor blades to dip, injuring or killing anyone who is not positioned low enough to avoid the blades.
- If you are wearing a helmet, tighten the chinstrap. No other hats should be worn around the LZ.
- Loose objects, such as stethoscopes, should be secured in a pocket or worn securely around your neck, so that rotor wash will not cause them to flap or blow away.

Danger

This phenomenon can occur while the rotor RPM speed up or slow down. A sudden wind gust can cause the rotor blades to dip, injuring or killing anyone who is not positioned low enough to avoid the blades.

Hand Signals

Figure 5-14 illustrates basic hand signals that can help you assist a pilot in completing a safe landing. These include:

- Extend both arms and point to indicate the landing zone.
- Cross and uncross your arms (like the letter "X") above your head to wave off landing. This signals danger and that the pilot should immediately abort the landing.

The landing zone may be marked by small, weighted orange cones, strobe lights, or flares at each corner. Spreading a ground cover of contrasting color may enable the pilot to better find the LZ. This must be removed before permitting the helicopter to land.

Night-Flight Operations

Obviously, it is more difficult for a helicopter to land in an emergency landing zone at night than during the day. When helping a helicopter land at night, use flashlight wands so that your hand signals can be readily seen by the pilot. Night-flight operations usually require a larger landing zone than those conducted in daylight. The safest way to mark a landing zone at night is with a flameless light source, such as single strobe light or a road flare, at each corner of the zone. (Keep in mind, however, that flares pose an ignition hazard, especially if they are blown about by rotor wash.) The rotating beacon lights on some emergency vehicles also can help pilots find landing zones, both during the day and at night.

As for other lights, all nonessential lights should be turned off, especially during actual takeoffs and landings. Vehicle lights used to illuminate the landing zone should be on low beam. During a helicopter's approach, landing, and takeoff, all spotlights, floodlights, television camera lights, photoflashes, and handheld flashlights should be pointed at the ground, *never* at the helicopter or in the pilot's face, as this can ruin the pilot's night vision, temporarily blinding him or her. (Red or blue lights, such as flashlight wand covers, do not hinder night vision.) Maximum safety is ensured by keeping unauthorized personnel, news media, and bystanders at least 200 feet from the helicopter. In addition, ensure that all light markers, flags, and streamers are secured against rotor wash.

Surface Wind Direction

surface wind
aircraft land and take off into the wind, so the pilot must know the direction (and speed, if possible) of the prevailing surface winds

A common cause of helicopter accidents involves inaccurate or unavailable information on wind direction and speed at landing sites. Aircraft land and take off into the wind, so the pilot must know the direction (and speed, if possible) of the prevailing **surface winds**. During daylight, wind direction can be indicated simply by placing a highly visible flag or streamer upwind and just outside of the

Figure 5-14 *The proper helicopter hand signals.*

LZ perimeter. At night, indicate wind direction by placing an extra strobe or flameless light stick just *inside* the perimeter on the *upwind* side of the LZ boundary. As always, everyone should be cautious of rotor wash.

STRATEGIES FOR PLANNING AHEAD

As you develop your Standard Operating Guide for aircraft accidents, create checklists to help you ensure that once at an accident site, you do not forget anything. Also, it is helpful to arrange airport emergency exercises and to preplan how to care for your crew during grueling missions.

Developing a Tactical Checklist

As has been noted previously, your training should include aircraft familiarization walk-through exercises at your nearby airport or military airfield. Understanding the different characteristics of each aircraft type helps you define and plan tactical operations, including formulating a comprehensive tactical checklist such as this one. (An aircrew member or mechanic should supervise any aircraft familiarization training.) This is *your* checklist. Add any topics or items you think are important.

Tactical Checklist

Cockpit

- ❑ Cockpit and crewmember seating locations
- ❑ Oxygen and other pressure cylinders
- ❑ Safetying ejection seats, canopy jettison, and explosive hatches by personnel aware of the proper procedures and hazards associated with these tasks
- ❑ If it is a Private (General Aviation Aircraft), confirm/rule out if it is equipped with a Ballistic Recovery System (BRS)

Ballistic Recovery System (BRS) (*if applicable*)

- ❑ How to disconnect/release restraints belts, harnesses, and communication wires attached to crewmembers' helmets
- ❑ How to shut down aircraft engines and the auxiliary power unit (APU). This includes throttles, engine T handles, aircraft fire-extinguishing systems, condition levers, and magnetos (in propeller aircraft)
- ❑ Location of battery shutoff and/or master switch

(*continued*)

Tactical Checklist (*Continued*)

- ❑ Location of emergency signal flares
- ❑ Location and operation of emergency escape hatches and window exits
- ❑ Location of oxygen shutoff controls
- ❑ Location of walk around oxygen bottles location of hand-held fire extinguishers
- ❑ Any other hazards or concerns within this area

Main Fuselage/Cabin/Cargo Compartment: Interior

- ❑ Location and operation of all emergency exit hatches and doors (from inside and outside)
- ❑ Deployment of emergency slides
- ❑ Location and operation of all normal entry and exit doors
- ❑ Location of any emergency signal flares
- ❑ Location of emergency oxygen cylinders, generators, and walk-around cylinders
- ❑ Location of any additional fluid reservoirs and hydraulic lines (if known)
- ❑ Crew and passenger locations
- ❑ Emergency cut-in areas (inside view)
- ❑ Location of oxygen shutoff controls
- ❑ Location of walk-around oxygen bottles location of hand-held fire extinguishers
- ❑ Any other hazards or concerns within this area

Aircraft Exterior (as applicable based on aircraft type)

- ❑ Danger zones
- ❑ Location and operation of weapons and ordinance
- ❑ Location of danger zones associated with flare and chaff dispensers (military and some commercial aircraft)
- ❑ Location and operation of all emergency canopy jettison controls
- ❑ Fuel tank locations and fuel capacity
- ❑ Fuel system vents/dumping ports
- ❑ Location and operation of normal entry doors
- ❑ Location and operation of cargo doors
- ❑ Hydraulic tanks and capacities
- ❑ Fire access panels on engines and elsewhere
- ❑ Cut-in areas for rescue of aircrew and passengers
- ❑ Battery location and quick disconnect
- ❑ Location of aircraft fire-extinguishing agent tanks, type of agent.
- ❑ Location of composite materials (if known)
- ❑ Other hazards not listed above

Examples of tactical checklists are included in **Figures 5-15; 5-16a, b;** and **5-17.**

Airport Emergency Exercises

Any airport with regularly scheduled passenger flights should review their airport emergency plan at least once every 12 months. This review must include all

ANYWHERE FIRE AND EMERGENCY SERVICES

AIRCRAFT TACTICAL COMMAND WORKSHEET

SITUATION FOUND HIGH/MED/LOW IMPACT

A/C TYPE: ____

FUEL REMAINING: ____

TAIL #: ____

KNOWN CASUALTIES: ____

PROBLEM: ____

SECTOR 3 SECTOR 4

SECTOR 2 SECTOR 1

MAJOR CRASH REMINDERS:

RESCUE: ____ CORDON: ____ FIRE ATTACK: ____

HAZMAT: ____ MUTUAL AID: ____ RESUPPLY: ____

WINDS: ____

KNOTS: ____

TEMP: ____

315 N 045

270 – W E – 90

225 S 135

MUNITIONS/HAZ. CARGO

FIRE SYMBOL: ____

TYPE: ____

QTY: ____

HOT BRAKES

START: ____

SAFE: ____

COMMAND ASSIGNMENTS

INTERIOR: ____

EXTERIOR: ____

RESCUE: ____

VENTILATION: ____

RESUPPLY: ____

OTHER: ____

TACTICAL BENCHMARKS

ALL CLEAR: PRI. SEC. FIRE CONTROL: ____

LOSS STOPPED: ____ EMERGENCY TRAFFIC: ____

CASUALTY COUNT (DIM)

DEAD: ____ INJURED: ____ MISSING: ____

FIREFIGHTER ACCOUNTABILITY

INTERIOR: ____ EXTERIOR: ____ REHAB: ____ RECALL: ____

Figure 5-15 *A tactical worksheet. (Courtesy of the U.S. Air Force.)*

airport emergency plan exercise
drills used to evaluate the preparedness of an airport's emergency response teams and of surrounding municipal agencies in the event of a major disaster

agencies that may be called on to respond in accordance with this plan. There should be a full-scale **airport emergency plan exercise** at least once every three years.

Contact airport managers or other people in charge of the facility to schedule a tour, including hangars and other important buildings. Also, contact pilots or mechanics to arrange aircraft familiarization walk-throughs. These people are

WORKSHEET

TABLE 3.7-3. AIRCRAFT MISHAP HAZARDS ASSESSMENT (TABLE 1 OF 2)

MISHAP/DATE: ______________________

AIRCRAFT MDS: ______________________

LOCATION: ______________________

CATEGORY	HAZARD	CONDITION	LOCATION
FIRE	FUEL AND FUEL TANK		
	OVERHEATED BATTERY		
	HYDRAULIC FLUID		
	LEAKING OXYGEN		
	LEAKING OR HOT BATTERIES		
	SMOLDERING MATERIALS		
	SMOKING ON SITE		
	CUTTING TOOLS AND OTHER HEAT GENERATING SOURCES		
	SIGNAL FLARES		
	ANTI-ICING FLUID		
	HOT BRAKES AND TIRES		
HIGH PRESSURE SYSTEM	HYDRAULIC ACCUMULATORS		
	HYDRAULIC LINES		
	PNEUMATIC SYSTEMS		
	SHOCKS, STRUTS		
EXPLOSIVE	HOT BRAKES AND TIRE		
	UNSPENT MUNITIONS AND WARHEAD		
	AIRCRAFT ENGINE FIRE BOTTLES (APU)		
	EJECTION SEAT OXYGEN BOTTLES		
	LIQUID OXYGEN BOTTLES		
	PYLON EJECTOR CARTRIDGES		
	CANOPY AND EJECTION SEAT DEVICES. DROGUE GUN		
	EXTERNAL FUEL TANK EJECTOR CARTRIDGE		
ELECTRICAL	POWER LINES		
	LIVE WIRES		
RADIOLOGICAL	LANTRIN POD		
	NUCLEAR WEAPONS		

Figure 5-16(a) *Tactical worksheets. (Courtesy of the U.S. Air Force.)*

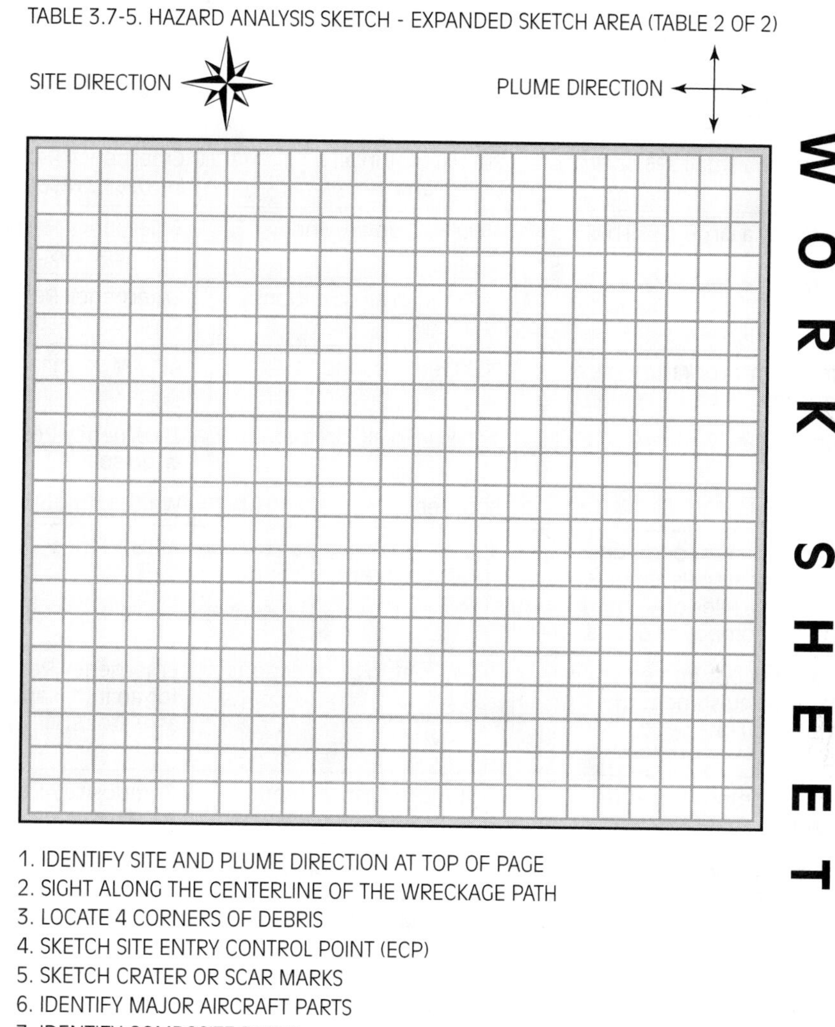

1. IDENTIFY SITE AND PLUME DIRECTION AT TOP OF PAGE
2. SIGHT ALONG THE CENTERLINE OF THE WRECKAGE PATH
3. LOCATE 4 CORNERS OF DEBRIS
4. SKETCH SITE ENTRY CONTROL POINT (ECP)
5. SKETCH CRATER OR SCAR MARKS
6. IDENTIFY MAJOR AIRCRAFT PARTS
7. IDENTIFY COMPOSITE PARTS
8. LOCATE HAZARDS: SEE CHECKLIST
9. SKETCH BLOODBORNE PATHOGEN AREA

ACCURACY IS NOT NECESSARY.!!!

Figure 5-16(b)(*contd.*)

happy to assist you in learning about the various aspects of their assigned airplanes. (You might even place these individuals on your response plan as technical consultants in the event of an aircraft accident.) Another invaluable source of information about various aircraft is the engineering diagrams and data readily available on the Internet and from your airport fire departments and aircraft manufacturers.

Table 3.7-1 Cordons (Street 1 of 2)

SCENARIO	INITIAL ISOLATION ZONE[1]	RESOURCE
Initial approach for a small JP8-spill	80–160 feet in all directions	Emergency Response Guidebook for a kerosene response.
Initial approach for a large JP-8 spill	1000 feet downwind	Emergency Response Guidebook for a kerosene response.[2]
JP-8 fire	1/2 mile in all directions	Emergency Response Guidebook for a kerosene response.[2]
Any fires burning in wreckage where HAZMAT is present	2000 feet upwind	AFPAM 91-211
Unknown substances	1/2 mile in all directions	Emergency Response Guidebook for a large spill.[2]
B-2 mishap	5000 feet	Whiteman Oplan 32-1
Helicopter approach for conditions: a. Hazardous material spills. Smoldering composite debris b. Piles of burnt and/or shattered composite debris.	500 feet above ground level 1000 feet horizontally	AFI 32 Series
Burnt composite debris immediately following the extinguishment of flaming combustion or smoldering.	30–80 feet in all directions	Emergency Response Guidebook for an initial approach for an asbestos spill.[2]
Burnt or broken composite debris piles - work zone	Not more than 25 feet	Composite Aircraft Mishap Safety and Health Guidelines.[3]

[1]The on-scene-commander will determine necessary cordon distances.
[2]U.S. Department of Transportation. 2000 North American Emergency Response Guidebook. A guidebook for First Responders During the Initial Phase of a Hazardous Materials/Dangerous Goods Incident.
[3]Air Standardization Coordinating Committee, Advisory Publication 25/41.

Figure 5-17 *Suggested cordon and isolation areas. (Courtesy of the U.S. Air Force.)*

As noted in previous chapters, many agencies are required to work together during an aircraft incident. To familiarize all of these agencies with each other's incident procedures, conduct frequent workshops involving airport infrastructure, airline, law enforcement, fire, medical, volunteer organization, and media personnel. In addition, all fire departments (military and civilian) should ensure that a common radio frequency is available on their mobile communications equipment and communications centers.

Military and civilian fire departments often need to work together during an aircraft accident. In the past, this was not always an easily accomplished goal, because some military fire chiefs were interested only in what occurred within the confines of their base, and some civilian fire chiefs excluded military fire agencies from joint training exercises. Today, however, civilian and military fire departments in many communities train side by side and include each other in their response plans. (Do not be surprised, however, if information about certain military aircraft, weapons, or buildings is classified.)

CASE STUDY

A case in point regarding the need for cooperation between military and civilian fire departments involved a traffic accident with a gasoline tanker truck. The accident occurred late at night in a suburban area. Thousands of gallons of gasoline were leaking from the damaged tanker and running into the street's storm drains.

An Air Force base was nearby, but two of its ARFF vehicles were out of service and inoperable, and the base was unable to respond to the incident. The community resented this, despite the base's historically excellent working relationship with community fire departments. For security reasons, base personnel were not permitted to explain to the community why they were unable to respond to the tanker accident. If a greater degree of communication had been allowed, the community might have understood what had happened and not felt such animosity toward the military personnel.

Training with other fire departments, including airport, military, or industrial fire brigades, is important. You may be called on to respond to a fire at a military base, an airport, or a large industrial complex. Preplanning with all other agencies who may be involved in an incident prevents confusion and promotes a safe, effective operation. Tabletop training exercises are a great means to develop and practice tactical procedures. During these exercises, you can make mistakes and repeat the exercise as many times as needed or desired without tying up response equipment or exhausting crews.

Some training providers have invested much time, effort, and money into designing impressively detailed and elaborate simulators and materials. These simulations are expensive, but they are well worth the money spent: Don't let a limited budget exclude you from this excellent training method. Whenever possible, however, try to arrange hands-on training exercises in addition to tabletop or simulator training. **Figure 5-18** shows county firefighters and airport fire department personnel training together and practicing how to extinguish a fire involving burning aircraft tires on this aircraft training mockup.

Figure 5-18 *Structural firefighters training with airport firefighters.*

Another helpful took is airports' response checklists, which may mirror many items on your own list. Please refer to Appendix B for sample checklists for pre- and post-incident response. Understand that these are only *guidelines* used by other agencies and that these agencies assume no liability for your actions if you decide to follow these suggestions.

Aircraft Accident Emergency Resource Checklist

Operational Needs

- ❑ Dispatch of response equipment for first, second, or additional assignments
- ❑ Mutual aid from other fire agencies (include the nearest airport/airbase fire department)
- ❑ Is their response feasible?
- ❑ Hazardous materials teams
- ❑ Environmental management

- ❑ Federal Aviation Administration (FAA): 1-202-267-3333
- ❑ National Transportation Safety Board (NTSB): 1-202-314-6290
- ❑ Availability of foam
- ❑ Nearby airports and their resources
- ❑ Military
- ❑ Refineries
- ❑ Other fire agencies (You may have to specify exactly what you need from them.)
- ❑ Food and water for response personnel
- ❑ Manpower sufficient? Reevaluate and update on an ongoing basis
- ❑ Foam equipment
- ❑ Water tenders
- ❑ Pumpers/Medic units
- ❑ Heavy duty or specialized rescue squads
- ❑ Hazardous materials teams
- ❑ Cell phones
- ❑ Additional radio communication equipment
- ❑ Portable cooling fans/refrigeration units
- ❑ Sheet plastic
- ❑ Busses to transport survivors
- ❑ Critical stress debriefing personnel, grief counselors
- ❑ Blankets
- ❑ Sand
- ❑ Dump trucks
- ❑ A specialized tanker truck that can remove the remaining fuel if the site contains residual fuel

Aircraft Accident in Water

- ❑ Boats
- ❑ Life preservers for all responders working in proximity of water
- ❑ To retrieve human remains:
  - ❑ Pool skimmers
  - ❑ Five-gallon plastic buckets with lids
  - ❑ 50-gallon plastic drums (with water tight lids)
  - ❑ (4) One-gallon bottles of bleach for decontaminating equipment exposed to human body fluids.
  - ❑ Large plastic bags
- ❑ Anything else you feel is needed to meet your response objectives

Airports can be tremendous resources for unusual equipment, such as the mobile air-conditioning unit shown in **Figure 5-19** that normally is used for cooling aircraft while they are parked. This is an example of the resources and equipment that can be provided by local civil and military airports.

If an aircraft has crashed into a lake, river, or the ocean, or other body of water, your personnel will need the following:

- Personal flotation devices for all responders
- Boats
- Swimming pool skimmers (with attached poles) for retrieval of personal effects, aircraft debris, and human remains
- Containment booms to manage drifting wreckage, and leaking fuel and oil leaking from the aircraft.
- Body bags
  - Plastic bags, plastic buckets, cans, and barrels (biological containment)
  - Labeling tags that withstand exposure to weather and water

Who do you have that can provide that service quickly and effectively? Don't forget portable heating and cooling, portable light units, portable electric

Figure 5-19 *Airport mobile air-conditioning equipment.*

generators, and aircraft crash recovery equipment. Local merchants may provide other items if local government resources are exhausted or not available.

These important things local merchants can provide would include but not be limited to:

- portable fans and heaters
- electric generators and lighting
- water misting units to cool your people in hot weather
- folding chairs, lawn furniture
- meals
- fruit juices
- water
- gasoline-powered electric generators
- sheet plastic
- portable tents
- foods, drinks, water, ice
- portable toilets
- a fuel truck to supply fire vehicles or aircraft

Caring for Your Crews

When your agency responds to a large aircraft accident, your crew is pushed to its limits, both physically and emotionally. Monitor your personnel for signs of physical fatigue, heat exhaustion, and psychological stress.

Note
To help your crews prepare themselves emotionally to face a crash scene, conduct a thorough pre-entry briefing informing responders what they are likely to see, hear, and smell.

To help your crews prepare themselves emotionally to face a crash scene, conduct a thorough pre-entry briefing informing responders what they are likely to see, hear, and smell. (If possible, request volunteers for the more gruesome tasks, such as gathering human remains, and whenever possible, select the older rather than the young volunteers. Bluntly, this is because a younger volunteer faces a longer life of bad memories than does an older volunteer.)

To help rescuers sustain their physical strength, ensure timely and realistic crew rotation. Personnel in the debris field, especially when they are working in extreme climatic conditions or performing the wrenching tasks of dealing with the wounded and dead, must be relieved and sent to a rehabilitation area (described in the next paragraph). Advise rescuers and volunteers to request rotation out of an area if you or they feel that they have seen enough for a while.

Provide a rehabilitation (rehab) area for your people that is out of the public eye, away from news cameras (let the public information officers do their job and deal with the media), and to which access is limited to authorized personnel. This area can be located inside a nearby building or behind a large truck, where tarps are hung to create privacy walls. The rehab area should offer physical assessments and medical and psychological treatment. The latter should include the services of chaplains and stress debriefing counselors. Local chapters of organizations such as the American Red Cross and the Salvation Army are available to care for your people as well as passengers, their families, and people in a crash area who may have been displaced from their homes.

In many states, the incident commander is accountable for safety during the *entire operation,* including contractor operations during cleanup and disposal. The IC *must* conduct an in-person, face-to-face briefing with the next person who succeeds the IC in assuming control of the scene. This person may be NTSB personnel, a military investigator, an environmental authority (often after rescue crews have been released), the coroner (called Mortuary Affairs by the military), or another authority. In a nonmilitary situation, a contractor may perform debris cleanup. This is likely to happen long after any emergency **mitigation** is complete.

mitigation
activities designed to reduce or eliminate risks to persons or property or to lessen the actual or potential effects or consequences of an incident

SUMMARY

This chapter addressed the following typical response tactics and strategies:

- Your role as the answer to someone's cry for help
- What you may face at a crash scene
- Being overwhelmed by the initial shock of encountering an aircraft accident
- Approaching the incident
- Assisting airport fire departments
- Mass casualty incidents
- Services you may be required to provide
- Helicopter (medevac) operations
- Using checklists

KEY TERMS

Aeromedical (medevac) operations Emergency aircraft (usually helicopter) operations that require teamwork among different agencies, including medevac aircrews, fire, police, ground-based EMS personnel, or military agencies. When helicopter crews are transferring patients, evacuees, or emergency officials, ground assistance is required to select a landing zone (LZ) that is safe and close to the personnel requiring transport.

Airport emergency plan exercise The Federal Aviation Administration (FAA) requires a full on-site aircraft disaster exercise once every three years (thus, this is often referred to as a "tri-annual"). These drills are used to evaluate the preparedness of an airport's emergency response teams and of surrounding municipal agencies in the event of a major disaster. Numerous local agencies are encouraged to participate, including ambulance services, EMS authorities, hospital trauma centers, the local American Red Cross, and officials from the Transportation Security Administration and the FAA.

Aqueous film forming foam (AFFF) Fire-extinguishing agent that contains fluorocarbon surfactants and spreads a protective blanket of foam that extinguishes liquid hydrocarbon fuel fires by forming a self-sealing barrier between the fire and fire-sustaining oxygen. Note: Application of foam % means the percentage of concentrate in the final solution (e.g., 6% means 6% foam and 94% water).

Augmentee A person who volunteers, or is recruited, to assist in an emergency operation. These people can be utilized for nonhazardous duties, such as assisting with crowd control, obtaining ice, assisting in cutting wood for cribbing, and so on.

Body substance isolation (BSI) This term refers to wearing personal protective equipment (PPE), such as gloves, goggles, gowns, and masks, while rendering first aid or in any other way risking exposure to any human body fluids.

Class A foam This firefighting foam is mixed with water, allowing the extinguishing agent to penetrate deep-seated fires involving matted grasses or other dry vegetation, bales of cotton, cardboard, or other Class A combustibles requiring a penetrating fire-extinguishing agent. It is also used to temporarily coat exposed flammable vegetation and structures, affording temporary protection from fire.

Class B foam This is used to control and extinguish fuel fires (Class B fires), such as those caused by gasoline or other liquid hydrocarbon fuels.

Dry chemical A fire-extinguishing agent used to extinguish Class A, B, or C fires and composed of ammonium phosphate (ABC), potassium bicarbonate (brand name, Purple K), or similar material.

Dry powder A fire-extinguishing agent used for Class D (metal) fires and usually made of powdered graphite or similar materials. Brand names include Metyl-X™.

Emergency escape route This applies to a safe escape path for passengers and crew of an aircraft crash landing to quickly escape with minimal chances of burns or sustaining secondary injuries as a result of escaping the aircraft. For firefighters or other emergency responders it is a designated escape route for them to exit a "danger zone" in the event it becomes too dangerous to continue their involvement in the operation.

Fire-extinguishing agent The material used to extinguish a fire, such as Class B foam, Class D dry powder, water fog, hose streams, gaseous agents (e.g., carbon dioxide or Halon), or a combined dry chemical-water application.

Fixed-base operator (FBO) An entity at an airport, which renders maintenance, storage, and servicing of aircraft. They may also rent aircraft to licensed pilots.

Landing zone (LZ) An area designated for the transfer of medical patients or evacuees from a ground area to board an aircraft, usually a helicopter. The landing zone may be a roadway, school, parking lot, or open field. The surrounding area should be clear of street lamps, trees, power poles, buildings, fences, wires, radio towers, or other obstacles.

Level of impact This refers to the speed of the aircraft when an aircraft struck the ground:

High-impact crash Characterized by extensive break-up and disintegration of the airplane. Most of the wreckage will be in small pieces. It is likely there will be a crater at the impact site. It is almost impossible for an aircraft occupant to survive a high impact crash.

Medium-impact crash Characterized by the aircraft fuselage breaking apart into several large pieces. The likelihood of a post-crash fire is great. There is a good chance of survivors, but they are likely to have sustained moderate to severe trauma.

Low-impact crash Can be either on or off an airport runway such as a county road or open field. The fuselage is basically intact, and it is common for people to survive this type of crash. Most likely you will encounter injured people and fires on the ground as well as inside the aircraft.

Mitigation This term describes activities designed to reduce or eliminate risks to persons or property or to lessen the actual or potential effects or consequences of an incident. Mitigation measures may be implemented prior to, during, or after an incident. Mitigation involves ongoing actions to reduce exposure to, probability of, or potential loss from hazards. Fire suppression, rescue, and managing hazardous materials release may fall under this descriptive term.

National Defense Area (NDA) The NDA is an area established on nonfederal lands located within the United States or its possessions or territories. The purpose is for safeguarding classified defense information or protecting Department of Defense (DOD) equipment and/or material by keeping non-authorized people away from such sensitive materials. It becomes *temporarily* in control of the DOD.

Occupational exposure This term is associated with universal precautions and refers to unprotected contact with potentially infectious materials that may occur during the performance of an employee's duties.

Phonetic alphabet A spoken-word alphabet designed to eliminate confusion for listeners when letters sound alike.

A – Alpha	J – Juliet	S – Sierra
B – Bravo	K – Kilo	T – Tango
C – Charlie	L – Lima	U – Umbrella
D – Delta	M – Mike	V – Victor
E – Echo	N – November	W – Whiskey
F – Foxtrot	O – Oscar	X – X-ray
G – Golf	P – Papa	Y – Yankee
H – Hotel	Q – Quebec	Z – Zebra
I – India	R – Romeo	

Raindrop effect The most efficient aqueous film forming foam (AFFF) firefighting technique. The "raindrops" do not form when leaving the nozzle or turret, but when they contact the burning fuel. The stream must be adjusted so it is not so sparse that it evaporates or is carried away by heat updrafts from the fire.

Rotor Rotating airfoil assembly (propeller) of helicopters.

Surface wind A common cause of helicopter accidents involves inaccurate or unavailable information on surface wind direction and speed at landing sites. Aircraft land and take off into the wind, so the pilot must know the direction (and speed, if possible) of the prevailing surface winds.

Tactics Tactics is the "doing" of all the necessary operational activities needed to accomplish the goals set by the IC. Tactics must follow the goals and objectives set forth by the strategic plan.

Tilt-rotor aircraft Called "powered-lift vehicles," these vehicles can take off, land, and hover similarly to a helicopter, and then can change wing angle and fly like a turboprop airplane.

Triage This term means "to sort out." It is used in emergency medicine, including by medical or other emergency responders, especially during mass-casualty incidents and in crowded hospital emergency rooms. It is the process of sorting patients for priority treatment based on factors such as severity of injury, likelihood of survival, and available resources. Systems of designation include ranging systems (from "good to grave" or "1 to 5") and color codes; use whichever system is common to your area. The START (simple triage and rapid treatment) system is very popular and is taught by the American Red Cross and EMS teaching institutions.

Universal precautions These work practices comply with OSHA regulations and includes wearing body substance isolation equipment and clothing. Universal precautions include engineering controls (built-in protection), wearing personal protective equipment (PPE), decontamination, and properly disposing contaminated materials.

REVIEW QUESTIONS

1. List the necessary steps in following universal precautions.
2. You are on scene with another pumper (engine company) where a twin engine General Aviation plane has sustained a low impact crash. The left wing is on fire. Occupants are trying to escape. What is your most important tactic?
3. Describe the predicted clues that would indicate a high impact crash. What is the likelihood of survivors?
4. Bystanders are at the scene of an aircraft accident and are getting in the way of rescue operations. How, with limited manpower, can you keep the crash area secure for an investigation?
5. Personnel are being decontaminated after exposure to smoke and burned composite particulate. Why should hot or very warm water be avoided?
6. You are approaching an F-14 that has made a successful "wheels up" emergency landing in a muddy pasture. The aircraft has missiles attached to the wings. What danger do these missiles pose from the rear?
7. You may encounter crewmembers that safely ejected or "bailed out" of the stricken aircraft. Because of the intense forces of the ejection seat and the accompanying blast of wind, what injuries may these people have suffered?
8. Why do we approach or leave the helicopter from downhill in full view of the pilot between the 10 and 2 o'clock positions of the front of the aircraft?
9. When practicing universal precautions, what must we wear to ensure appropriate protection from disease carrying bodily fluids?
10. Where is the unusual location for the main entry or exit door on the majority of all airplanes that are designed to carry freight and passengers?
11. Driving a vehicle fast through smoke a dangerous idea at any fire, especially at an aircraft crash site?
12. What approach angle to a fuselage is safest for any weapons-carrying military aircraft?
13. What does the abbreviation *BSI* stand for in the context of universal precautions?
14. What can happen when water is directed at burning magnesium?
15. You have arrived and assumed command at the scene of a military airplane crash. How can you preserve national assets and security?

STUDENT EXERCISE

Figure 5-20 is a copy of the cover sheet/index for a major international airport. This emergency plan fits in with local, state, and the National Incident Management System. Study the topics of this plan.

- How would your agency fit in with a similar plan for your local airports?
- What special resources can your agency contribute to your nearby airport or military air field?

BASIC PLAN

Page

FUNCTIONAL ANNEXES

HAZARD SPECIFIC APPENDICES

Figure 5-20 *Sample emergency response plan (ERP)—table of contents.*

TERMINATING THE INCIDENT

Learning Objectives

Upon completion of this chapter, you should be able to:

- Understand how to preserve evidence of an accident, including encouraging witnesses to interface with the appropriate authorities, creating notes about and diagrams of the scene, and safeguarding data recording devices.
- Understand how to deal with the media.
- Understand what is involved in the termination and post-incident review phases of an accident, including the importance of the initial (informal) and formal post-incident critique/debriefs.
- Understand basic information about post-traumatic stress disorder, including recognizing the signs of it and how to minimize the likelihood of developing it.

INTRODUCTION

The old saying "the job isn't done until the paperwork is finished" is only partially true in the context of emergency response.

The paperwork is certainly important, because investigators must have access to as much information as possible to help them determine the cause of a crash. Witnesses should be encouraged to make statements to the proper authorities; notes and diagrams of an accident scene must be made and organized.

Emergency responders are not done when this is complete, however, because they are not doing just any job—every time they answer a call, they place themselves in harm's way and may have to witness others' horrific suffering.

Thus, although the termination and post-incident review phases of an aircraft accident are easily overlooked and often forgotten, they are tremendously important, because this is when responders can deal with, and receive help dealing with, trauma they may have suffered during an emergency response. Properly completing this phase of an incident can ensure that you and your crew are not devastated by the effects of extremely distressing incidents.

The post-incident review phase can last for days, weeks, months, or sometimes even years, depending on the size and complexity of an incident. The long-lasting effects of tragic experiences can be positive if the steps of the termination process are properly executed: Everyone involved can learn important lessons, about themselves, their coworkers, and what it means to do their job.

After any traumatic incident, look out for your crew: watch for changes from normal behavior and other signs of psychological distress. If you observe any, use the services of mental health professionals to help any responders who are experiencing long-lasting negative effects of what they witnessed.

OBTAINING ACCIDENT INFORMATION

The most important sources of information about an incident are witness statements, scene documentation, the data stored in an aircraft's black box, and physical evidence.

Witness Statements

At a crash site, people who have witnessed the accident may approach you stating they have information about the crash (which may include photos or video footage). Direct these people immediately to the designated **public information officer** (who is discussed in more detail later). If the witness information affects tactics—for example, if a witness reports seeing people trapped in buildings—it must be channeled to the proper people within your command structure. Such witnesses are valuable assets to investigation teams, who may find these statements useful in determining the causative factors of the accident.

public information officer
ensures that information is provided to the public by any means available, usually by the media, during an emergency

Documenting the Incident

It is important to write detailed notes about—and preferably take pictures of—an accident scene for many reasons. First, this information may offer clues to the aircraft's angle, speed, and force of impact. Second, if the aircraft's crash has produced craters or deep furrows in the earth, you need to know this so that you can avoid these dangers when approaching the site with apparatus. Third, this information can be provided to the media once it is cleared by security.

field notes
handwritten notes that include multiple observations of an aircraft incident, including the time, rough sketches of initial findings, and significant events; may be initially written in a simple notebook or paper, but should later be compiled in an incident journal

incident diagram
a sketch of the scene of an emergency incident, including aircraft, rail, highway, pipeline, maritime, HAZMAT, wildfire, structural fire, or other accidents or incidents

incident site safety plan
a drawing or diagram of an incident, such as a HAZMAT, accident, or other dangerous incident; must comply with safety procedures and include exclusion zones, division assignments, emergency escape routes, a decontamination corridor, and safe-refuge areas

Incident and Safety Notes You should produce **field notes**, and these notes should contain an **incident diagram**. This diagram can also act as your **incident site safety plan**. If you have time, personnel, and resources, designate a photographer to take pictures of wreckage and other significant data, to aid the investigation.

Field notes: These handwritten notes can be taken by different personnel who participate in the incident. They include multiple observations of an aircraft incident, including the time it occurred, statements from witnesses, rough sketches of initial findings, and significant events. During the initial phases of setting up command, these notes can be jotted down on paper or a check list, or written in a simple notebook, but they should later be compiled in an *incident journal,* which contains all of your information about an accident, including field notes, incident diagrams, and site safety plans. (The incident journal is discussed in more detail later.)

Incident diagram: This is a drawing of an accident site. The diagram should include information about fuel, cargo, and trapped personnel; it should note the locations of wreckage, bodies, leaking substances, cargo pallets, cargo cans, as well as the existence and location of craters, scratches to paved surfaces, gouges or skid marks in dirt, damaged buildings, and motor vehicles. Include responders' entry routes and initial actions, as well as variables such as climate, time of day, wind direction, and resources that were planned for but that are unavailable. When quickly sketching a diagram for your field notes, the diagram does not have to be pretty or to exact scale. (If the drawing is not to scale, however, this should be indicated.)

Figures 6-1 and **6-4** (see on pages 172 and 175) are examples of the handwritten incident diagram of an off-airport crash operation that used structural and water tender fire vehicles. Note that the diagrams show that factors such as wind and water resupply were considered. This incident was successfully controlled by a wildland fire department with the assistance of a nearby military base, which provided ARFF equipment. Note that the fire was suppressed using high-pressure water spray (also called fog) nozzle patterns, that resupply techniques were used, and that the ARFF and wildland fire departments worked effectively together.

Incident site safety plan: This plan must indicate the assorted zones of an aircraft crash or a hazardous materials incident. Incident site safety plans take the form of a diagram that shows the main dangers posed by an incident and the locations of the decontamination corridor, the safe-refuge area, support zones, and so on. The plan also should show the site's topography as it relates to

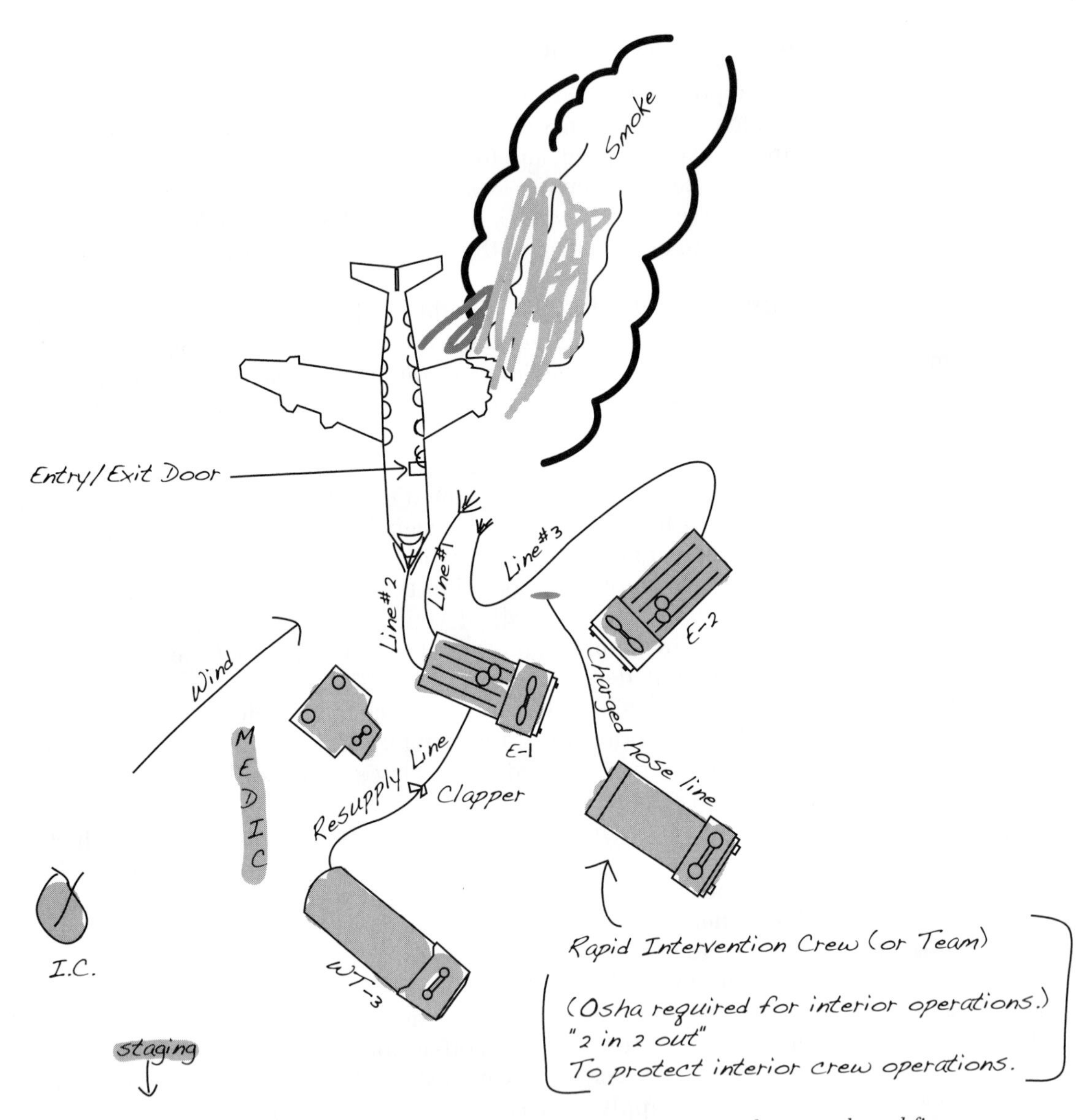

Figure 6-1 *Handwritten field notes: a tactical diagram of a private airplane crash and fire.*

drainage and firefighting foam or water runoff, as well as any other information needed for safety or tactical operations.

This plan is displayed during the pre-entry briefing. Obviously, time is critical when an aircraft is on fire, so, initially, the plan is hastily sketched in the incident commander's field notes. Later, a more precise plan can be created and included, with the first drawing, in the incident report. Note the grid paper in **Figure 6-2:** This template is preferred by many incident commanders for creating

TABLE 3.7-2 HAZARDS ANALYSIS SKETCH (SHEET 2 OF 2)

MISHAP/DATE: ______________________

AIRCRAFT MDS: ______________________

SITE LOCATION: ______________________

WEATHER DURING INCIDENT ____________

____

TERRAIN DESCRIPTION ____________

____

____

DIRECTIONS

SITE

DOWNWIND AND SPEED

COMMAND POST

IMPACT POINT

1. IDENTIFY SITE AND PLUME DIRECTION.
2. SIGHT ALONG THE CENTERLINE OF THE WRECKAGE PATH.
3. LOCATE 4 CORNERS OF DEBRIS.
4. SITE ENTRY CONTROL POINT (ECP).
5. SKETCH CRATER OR SCAR MARKS.
6. IDENTIFY MAJOR AIRCRAFT PARTS.
7. IDENTIFY COMPOSITE PARTS.
8. LOCATE HAZARDS: SEE TABLE 3.7-7
9. SKETCH BLOODBORNE PATHOGEN AREA

ACCURACY IS NOT NECESSARY!!!

LAND OR GROUND MISHAP

HIGH ANGLE HIGH SPEED IMPACT:

DEEP CRATER. WRECKAGE SPREAD IS SHORT AND WIDE.

HIGH SPEED LOW ANGLE IMPACT

SHALLOW CRATER. WRECKAGE IS LONG AND NARROW. HEAVY PARTS ARE FURTHEST AWAY.

LOW SPEED HIGH ANGLE IMPACT

WRECKAGE IS CENTRALIZED.

LOW SPEED LOW ANGLE IMPACT

GROUND SCARS LEADING TO WRECKAGE FRACTURE ALONG PRIMARY STRUCTURE.

AIRCRAFT DEBRIS MAY TUMBLE SEVERAL TIMES BEFORE REACHING ITS FINAL LOCATION AT THE SITE.

Figure 6-2 *A template of a formal diagram for aircraft accidents. (Courtesy of the U.S. Air Force.)*

the second, more formal drawing. These drawings are critical to an investigation, not only because they document invaluable information, but also because they are excellent memory joggers and help responders recall details of an operation. (*Note:* The template in Figure 6-2 was obtained from a public domain; you may reproduce it and use or modify it as needed.)

emergency locater transmitter (ELT)
this device is designed to automatically send a radio signal from a stricken aircraft in the event of a crash. Rescue agencies such as the military and Civil Air Patrol can locate a downed aircraft by tracking the signal from an ELT

Data Storage Devices

Figure 6-3 shows the location in a commercial jetliner fuselage of an emergency locater transmitter, a cockpit voice recorder, and a flight data recorder.

Most aircraft are equipped with the first item, an **emergency locater transmitter (ELT)** (shown at the extreme left in Figure 6-3). This device is designed to automatically send a radio signal from a stricken aircraft in the event of a crash. Rescue agencies, such as the military and civil air patrol, can determine the location of a downed aircraft by tracking its ELT signal.

cockpit voice recorder (CVR)
records conversation and other sounds inside the cockpit

The last two items, the aircraft's **cockpit voice recorder (CVR)** and flight data recorder (FDR), are crucial to accident investigation, so these devices must be located, photographed, and guarded where they are found (these should not be moved until investigators indicate that this is permissible). Do *not* permit

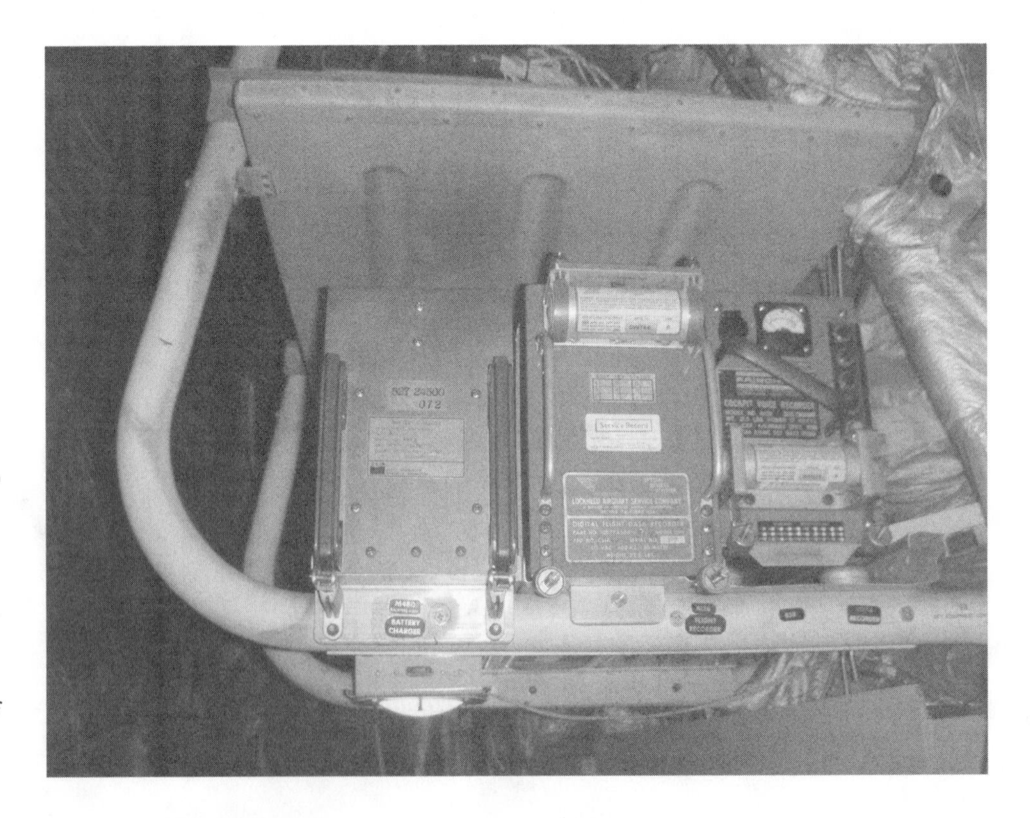

Figure 6-3 *An emergency locater transmitter (ELT), a cockpit voice recorder (CVR), and a flight data recorder (FDR), located in the tail of a large commercial jet airliner.*

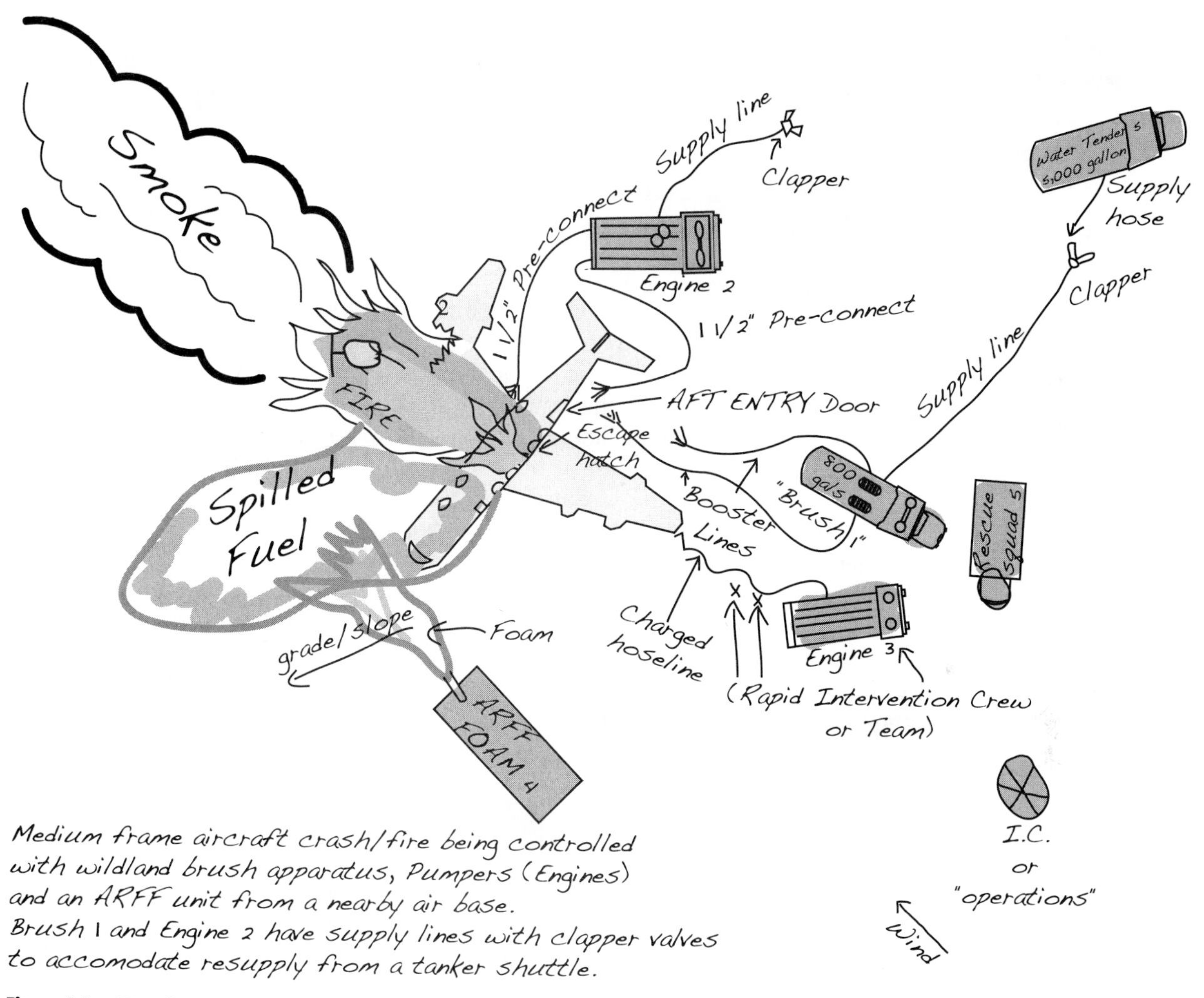

Figure 6-4 *Handwritten field notes: a tactical diagram of a commuter airplane crash and fire.*

anybody to handle, touch, or tamper with this equipment unless it is necessary to do so to prevent further damage to the equipment. CVRs and FDRs should be handed over only to the appropriate accident investigating authorities.

Remember from Chapter 2 that the term *black box* refers to the box that houses the cockpit voice recorder or flight data recorder (FDR). A CVR records crew conversations and sounds audible in the flight deck (cockpit). An FDR records in-flight information, such as speed, engine RPM, aircraft flight attitude, pitch, outside air temperature, vertical acceleration, and other variables. (FDRs may also be referred to as a *flight recorder* or *digital flight data recorder.*) Ironically, black boxes are painted bright orange to make them highly visible. These

■ Note
If these devices are found in water, then the appropriate officials (NTSB accident investigators) will place them in a bucket of the same water in which they were found.

devices usually are mounted in the tail section of an airplane. If these devices are found in water, then the appropriate officials (NTSB accident investigators) will place them in a bucket of the same water in which they were found.

WORKING WITH THE MEDIA

It is advisable to have a good working relationship with the media, because whenever an aircraft emergency occurs, it is likely that media organizations will send reporters to the scene. There are other, more important reasons to foster positive relationships with the media, however: They may obtain information useful to the accident investigation, and they can help you keep the public informed by releasing accurate and pertinent factual information. During an incident that requires evacuation, the media can assist in broadcasting information about the need for evacuation, evacuation routes, and other vital public safety information.

The incident commander should designate a public information officer (PIO) to function as the official point of contact with the media and to ensure that accurate information about an accident is provided to the public by any means available (usually by the media). PIOs provide the media, and thus the public, with appropriate information related to all phases of an incident. This information and its release should be coordinated with the incident commander and members of the unified command staff.

THE TERMINATION PHASE

The fire is out, and the causalities have been removed to medical facilities. Investigators are on scene, putting together the puzzle of how the accident occurred. This may be when you should terminate the emergency phase of the incident and pass the responsibilities to the next appropriate professional. If so, call your communications dispatch and advise them that you are terminating command. This may seem like the end of the incident for you, but in fact, this only begins a new phase of the accident.

Your Standard Operating Procedures may require you to create a detailed incident journal, mentioned earlier in this chapter. If so, this journal should contain information about what you encountered, other agencies that were involved in the incident, a roster of fire apparatus used, and the notation of other significant events (including their time), such as the arrival of personnel from co-responding agencies or specialized equipment. Your incident journal may begin with prepared check lists or with field notes.

Any pictures of wreckage should be duplicated and shared with the investigating agency in charge. If you have not already done so, draw an incident diagram that includes any pertinent information and place it in the incident journal.

Initial (Informal) Post-Incident Critique/Debrief

post-incident critique examines which parts of an operation went well and according to plan and which did not; used to identify operational shortcomings in a constructive manner and to design methods for overcoming those shortcomings

debriefing (psychological) gives individuals or a group the opportunity to talk about their experiences and how those experiences have affected them, and to discuss ways of dealing with their feelings about the incident

defusings limited to individuals directly involved in the incident; designed to quickly address responders' immediate needs

As soon as possible after an incident, and before anyone is released from the scene, you should have a group meeting—a **post-incident critique** or a critical-incident stress **debriefing**. This includes psychological and practical aspects, such as informing responders whether they may have been exposed to hazardous materials.

Psychological Debriefing For a debriefing to be effective, it is critical that incident participants are encouraged to talk about the event, because this is the beginning of the recovery process. A variety of methods can be used for debriefing. Some organizations follow a three-step program, but the most common program involves seven steps, which are

1. *Introduc*ing the intervener (facilitator) and establishing guidelines
2. Sharing *details of the event* from individual perspectives
3. Providing subjective *emotional responses*
4. Sharing *personal reaction and actions*
5. Noting any *symptoms* that personnel have exhibited since the event
6. Having the intervener *instruct* (assure) individuals that their responses to the event are normal
7. Having individuals *resume their duty* and return to their normal tasks

The debriefing includes incident commanders and other incident supervisors, who, along with the members of their crews, should share their perspectives on the events.

Regardless of which program is used, interveners watch responders to identify those who are not coping well and then offer these individuals additional assistance is at the conclusion of the process. Also, the final step of all these methodologies is *follow-up*; optimally, this occurs within 24 hours of the debriefing session.

Professionals in the field of critical-incident stress debriefing agree that talking about an incident is an effective means of reducing the impact of the stress responders experience during incidents. Reducing this stress, in turn, can minimize the incidence and degree of post-traumatic stress disorder, a serious form of clinical depression that can develop in responders (and survivors, of course) following a particularly devastating accident. (PTSD is discussed in more detail later in this chapter.) Debriefing facilitators should make every effort to reassure the responders that the job they just performed was an important part of the incident and that their actions concretely helped people in dire need. Interveners should give responders a well-deserved vote of confidence and recognize the hardships they endured in managing the scene of devastation.

Defusings are limited to individuals directly involved in the incident and are often done informally, sometimes at the scene. They are designed to quickly address responders' immediate needs.

employee assistance program (EAP) program offered by an employer to help employees address personal problems that might negatively affect their job performance or their physical and mental health

During the debriefing, advise employees how to contact grief counselors, chaplains, medical facilities, or anyone else who can offer assistance, if needed. Supervisors should be aware of whether their organization has an **employee assistance program (EAP)** and, if so, should provide responders with information about it. EAPs address such problems as alcohol abuse, drug abuse, financial or marriage problems, but a good EAP also provides employees with access to third parties—people who are not involved in the incident, but have knowledge of what happens at large-scale incidents—to help them cope with and solve issues related to the ramifications of an accident. Normally, EAPs include qualified counselors who have been trained to evaluate and assist in responders' recovery process.

At the conclusion of the debriefing, compile a roster including the names of everyone who was part of the operation, should they need to be contacted by accident investigators or organizational chaplains (as required), or for any other follow-up reasons, such as providing additional information about an incident.

Last, attempt to set a date to follow up with a thorough, formal post-incident debriefing or critique. This formal critique should include discussion of the particulars of the incident, and this information should be used positively to provide lessons learned from the incident.

Hazardous Materials Debriefing In addition to the psychological aspect of debriefing, responders should be advised of any hazardous materials they may have been exposed to, and incident commanders and facilitators should ensure that all people, clothing, and equipment has been decontaminated. Also, remind your crew to inform their immediate supervisor of any health concerns or symptoms that occur after the incident. Your Standard Organizational Procedures should include guidelines to follow if any responders experience either physical or psychological symptoms.

An employee exposure to hazardous materials is required by law to be documented and kept on file for at least 30 years after the employee has terminated employment or retired. This includes exposure to human bodily fluids, such as blood.

Formal Post-Incident Critique/Debriefing

Once all the paperwork has been completed, the reports have been written, and everyone has had a few days to think about the incident, a formal post-incident critique should be conducted.

Formal debriefings should take place within 72 hours of the incident. These debriefings are, in part, used for the same purpose as the initial post-incident debriefing: they give individuals or a group the opportunity to talk about the experience and how it affected them, and to discuss ways to deal with their feelings about the incident. The formal debriefing should be used, in part, as another opportunity to identify at-risk individuals and to inform them about the services available to them. These usually are the second level of intervention for those directly involved in the incident and the first intervention for those not directly involved. As with the first debriefing, the follow-up should be done with responders the day after the debriefing to ensure everyone is feeling well or, if necessary, to refer an employee to a professional counselor.

■ Note
Always begin and end post-incident critiques or debriefings on a positive note.

The post-incident critique of an operation is not intended to be a blame game. Always begin and end post-incident critiques or debriefings on a positive note. If you are the critique facilitator, do not let the debriefing turn into a finger-pointing and blame-assigning session. Such events are not productive and often do more damage than good.

Instead, these sessions should be used to recap the incident and as a medium to provide lessons learned and to determine ways to improve operations. Everyone should have an equal voice in the discussion.

Have someone take notes of the discussion, and use these notes to summarize the meeting. Keep in mind that these notes, positive and negative, should continue to be used after the critique; they should not be allowed to collect dust. Instead, apply the lessons learned to your training to help you and your crew improve your response process.

A last note about post-incident procedures: You must be aware of the proper procedures to follow if legal issues arise regarding an incident. If you or any of your employees are contacted by attorneys, you must notify your organization's legal department immediately. Any discussions or meetings with such attorneys must be coordinated and conducted by this department.

post-traumatic stress disorder (PTSD)
a specific set of emotional and psychological distress symptoms resulting from involvement with a stressful or traumatic event

critical incident stress management
a short-term assistance that focuses solely on an immediate and identifiable problem; the goal is to enable affected employees to return to their normal daily routines as soon as possible, while decreasing their chances of developing PTSD

MINIMIZING THE LIKELIHOOD OF POST-TRAUMATIC STRESS DISORDER

Your responsibility to your crew does not end with the completion of initial and formal post-incident critiques. In fact, depending on how your responders are able to deal with the effects of a traumatic incident, these debriefings may be only the beginning of your responsibility. Even responders with years of experience may need ongoing assistance with recovering from an incident, especially if they had never experienced the devastation of a large-scale or otherwise particularly horrific accident. Responders, whether inexperienced or not, may suffer long-lasting effects from terrible incidents, including **post-traumatic stress disorder (PTSD)**. PTSD can arise days, weeks, months, or even years after a traumatic incident. You and other supervisors need to be knowledgeable about and alert for the signs that a responder needs help.

Critical-incident stress management—i.e., management of the entire process from the time the incident occurs until the responder is able to adjust to what happened—mitigates the short- and long-term effects of incidents on responders by focusing solely on an immediate and identifiable problem. The goal of this process is to enable affected employees to return to their normal daily routines as soon as possible, while decreasing their chances of developing PTSD.

In addition to post-incident debriefings and follow-up care, *peer counseling* is vital to the recovery process. Holding in the anguish, frustration, or distress generated by an incident can lead to depression, PTSD, hypervigilance, and suicidal thoughts or acts. Everyone has to vent, and talking among one's peers is proven to be extremely effective in helping responders deal constructively with the emotional effects of traumatic incidents. In addition, such counseling is proven to alleviate mild to moderate cases of post-traumatic stress. (Severe cases of PTSD should be referred to a professional as soon as possible.)

Recognizing the Symptoms of Post-Traumatic Stress Disorder

During incidents, be alert for any behaviors that may indicate that your response personnel are becoming overwhelmed by the terrible things they are dealing with. (Refer to the information in chapters 4 and 5 about helping your crew manage the stress of traumatic scenes and caring for your crew.)

Ensure critical incident-stress-debriefing personnel have been dispatched to the scene. The American Red Cross, the Salvation Army, and clergy should be included in your response plan, so that you can contact these resources immediately to support your people. Follow your protocols for critical-incident stress debriefing, which, in addition to debriefings, should include post-incident follow-up interviews for all personnel involved in the incident.

After the incident, monitor your people for behavioral changes associated with depression, post-traumatic stress disorder, and other psychological problems. These symptoms include:

- insomnia
- sleeping too much
- nightmares
- weight loss
- weight gain
- alcohol or drug abuse
- feelings of helplessness
- hypervigilance (i.e., being in a continual state of extreme "fight-or-flight" alertness)
- obsessing about death
- inability to control anger
- excessive crying
- sexual dysfunction
- no emotional feelings at all (emotional numbness)

These symptoms indicate that a responder requires help from a trained medical and/or mental health professional.

FINAL THOUGHTS

Figure 6-5 shows the mass casualty phase of a large airport disaster exercise involving many agencies, such as the airport itself, the FAA, the NTSB, police, non-airport fire departments, emergency medical services, and disaster planners. If an incident of any magnitude happens in this community, these agencies will be well-prepared to deal effectively with it. This figure illustrates the core point we have tried to make with this personalized and simplified book: that knowledge, preplanning, preparation, and practice are your best tools for facing the crisis of an aircraft accident.

This book contains the basic information you need to consider when responding to an aircraft incident. We geared this information toward two audiences: responders who are not normally first in to such an emergency, but who may have to function in that role; as well as responders who normally are first on scene and must plan how to manage such incidents.

Remember, aircraft incidents, like other emergencies, almost *never* happen under ideal circumstances. When a crash occurs, you and your crew must rely

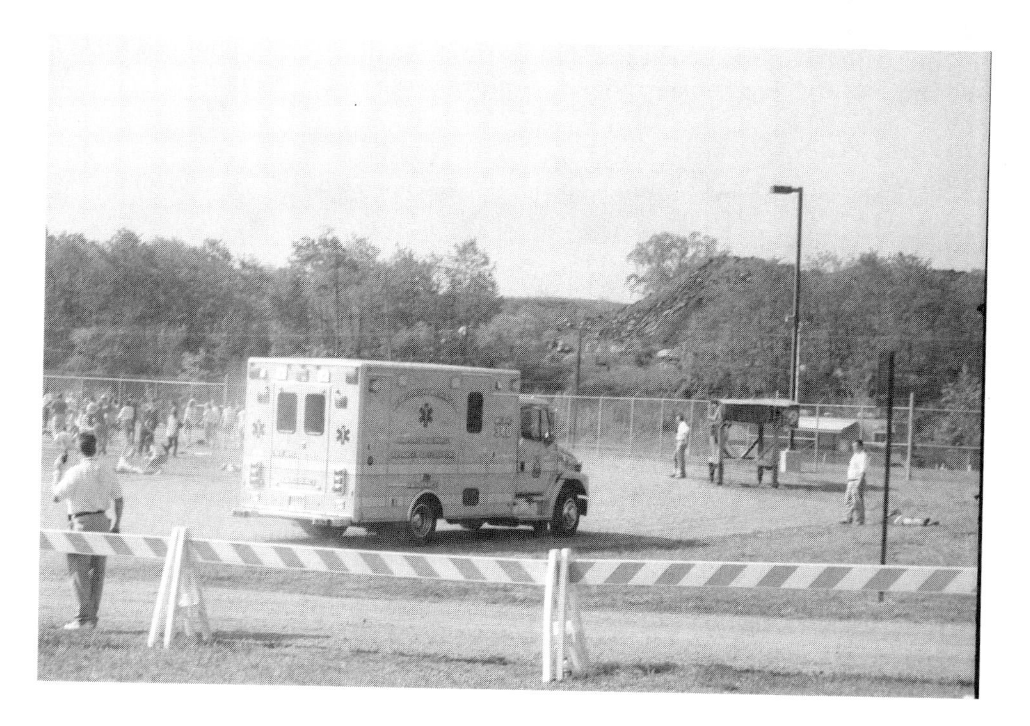

Figure 6-5 *An airport disaster exercise.*

on knowledge and expertise already gained. Thus, it is critical to thoroughly preplan, and make sure your plans are kept up to date and represent the most current information available. Practice, repractice, and perfect your abilities through hands-on training and tabletop exercises. Create checklists to help you focus on the tactical needs and requirements of an aircraft fire-suppression or rescue operation.

Finally, we encourage you to seek additional manuals and books on the subject to extend your knowledge of preparing to handle these incidents.

SUMMARY

This chapter focused on the final stages of an aircraft accident, the termination and post-incident review phases.

- Preserving clues about the cause of an aircraft accident is the responsibility of all responders at a scene. Take all possible steps to preserve evidence, including isolating the area from unnecessary personnel. In addition, ensure that wreckage from an aircraft, as well as skid marks; pavement scratches; and gouges, holes, or craters in the earth resulting from an aircraft's impact, remain undisturbed.
- Involvement in an aircraft accident can emotionally devastate responders.
- You and other supervisors must be alert for signs of physical and psychological distress, including depression and post-traumatic stress disorder, in your coworkers, and you must be familiar with your organization's employee assistance program.

KEY TERMS

Cockpit voice recorder (CVR) This device records conversation and other sounds inside the cockpit; it creates a continuous loop.

Critical incident stress management This short-term assistance focuses solely on an immediate and identifiable problem. Its goal is to enable affected employees to return to their normal daily routines as soon as possible, while decreasing their chances of developing PTSD.

Debriefing (psychological) A debriefing should take place within 72 hours of the incident. It gives individuals or a group the opportunity to talk about their experiences and how those experiences have affected them, and to discuss ways of dealing with their feelings about the incident. A debriefing usually is the second level of intervention for those directly affected by an incident, and often the first level of intervention for those not directly involved. During a debriefing, participants should be provided with information about the services available to them, and at-risk individuals should be identified. Post-debriefing follow-up should be done to ensure that everyone involved in an incident is feeling well or is referred to a professional counselor.

Defusings These are limited to individuals directly involved in the incident and are often done informally, sometimes at the scene. They are designed to quickly address responders' immediate needs.

Emergency locater transmitter (ELT) This device is designed to automatically send a radio signal from a stricken aircraft in the event of a crash. Rescue agencies such as the military and Civil Air Patrol can locate a downed aircraft by tracking the signal from an ELT.

Employee assistance program (EAP) A program offered by an employer to help employees address personal problems that might negatively affect their job performance or their physical and mental health. EAPs vary, but, ideally, they should address substance abuse, emotional distress, exaggerated safety concerns, physical health, dysfunction affecting marriage and family, death and grief, financial problems, legal problems, and other life-affecting events. These services usually are prepaid by employers, often through a third-party provider, and free to employees and their household members.

All information obtained during or generated by these programs is maintained in confidentiality. Employers usually are not informed of which employees are using EAP programs, unless there are extenuating circumstances that require an employer to be given this information. In that case, the proper release forms must be signed by the employee. In some cases, management may require employees to seek EAP assistance because their behavior or work performance has become unacceptable.

Field notes These handwritten notes include multiple observations of an aircraft incident, including the time, rough sketches of initial findings, and significant events. During the initial phases of setting up command, these notes may be written in a simple notebook or paper, but they should later be compiled in an incident journal.

Incident diagram This is a sketch of the scene of an emergency incident, including aircraft, rail, highway, pipeline, maritime, HAZMAT, wildfire, structural fire, or other accidents or incidents.

Incident site safety plan A site safety plan is a drawing or diagram of an incident, such as a HAZMAT, accident, or other dangerous incident. The plan must comply with safety procedures and includes exclusion zones, division assignments, emergency escape routes, a decontamination corridor, and safe-refuge areas. Before an entry is made into a HAZMAT, a safety briefing is conducted and all participants are advised of the action plan and of emergency safety contingencies specific to the incident area.

Post-incident critique This session examines which parts of an operation went well and according to plan and which did not: It is used to identify operational shortcomings in a constructive manner and to design methods for overcoming those shortcomings. Many agencies prefer to conduct post-incident critiques in a debriefing format. This focuses on addressing tactical/operational events and is different than a debriefing that is intended to address the fire-suppression/rescue-operation from a tactical standpoint.

Post-traumatic stress disorder (PTSD) This term refers to a specific set of emotional and psychological distress symptoms resulting from involvement with a stressful or traumatic event. For first responders, the most stressful events commonly are line-of-duty deaths, coworker suicide, multiple casualty incidents, and large-scale disasters. Responders who have experienced a traumatic event may develop PTSD if intervention by trained counselors is delayed.

Symptoms of PTSD may involve physical or psychological problems such as insomnia, sudden changes in weight (loss or gain), anxiety, depression, uncontrollable anger, withdrawal, lack of any emotion, nightmares and flashbacks, or hypervigilance and exaggerated "startle response."

Most people who experience traumatic events do not develop PTSD, which is primarily an anxiety disorder and should not be confused with the normal grief and adjustment that occur after traumatic events. For additional information, contact your local mental health association.

Public information officer During an emergency, the incident commander designates a public information officer (PIO) to ensure that information is provided to the public by any means available, usually by the media. Response organizations coordinate information with PIOs and clear all press releases with incident command before releasing information. PIOs function as the official point of contact for the media, and keep them advised of proper and accurate information related to all phases of an incident. This information should be in concert with the incident commander and members of the unified command staff. When responders are approached by the media, they should refer them to the designated PIO.

REVIEW QUESTIONS

1. Several months after a gruesome aircraft incident, your coworker seems withdrawn. He doesn't have his usual sense of humor and spends a lot more time listening to his portable stereo (headphones) than before. What might your friend be suffering from?
2. You were incident commander at a major aircraft accident involving hazardous materials. List all elements and information your incident journal should contain.
3. Describe the duties of a public information officer.
4. What should you do with any pictures of an aircraft crash taken by citizens or emergency responders?
5. If an aircraft's black box is found in water, what should you do with it?
6. What color is a black box (i.e., a flight data recorder or cockpit voice recorder)?
7. How can the media be an asset to an aircraft accident operation?
8. How long is an employer required by law to maintain records of an employee's exposure to hazardous materials or blood-borne pathogens?
9. You are performing your duties at an aircraft accident/crash site when you are approached by a television reporter who begins to ask you questions about the crash. What should you tell this reporter?
10. Web sites, including those containing aircraft rescue data, cease to exist, or the addresses may change. When this happens, how can you obtain current aircraft rescue information from the Internet?

STUDENT EXERCISE

Refer to the list of acronyms in the back of this book; these acronyms were taken from a state emergency response plan. After reviewing these:

- Add the acronyms for your state agencies.
- Place a check mark beside agencies that exist in your state ERP.
- Correct/change acronyms if they differ from those in your state ERP.
- Delete the acronyms for agencies that are not in your state ERP.
- Add telephone numbers for the agencies that are in your state ERP.

ADDITIONAL ACTIVITIES

Case Studies and Tactical Exercises

CASE STUDIES

As an emergency responder, you have the obligation to continue learning about the processes, methods, and industry of firefighting. This means staying current, for example, with developments in incident command, EMS, aircraft construction materials, and the configurations of airplanes. You can do this not only in the formal environments of classrooms, learning conferences, and field exercises, but also from the experiences of others. Presented in a nonjudgmental format, other responders' triumphs and shortcomings are of great value. There are many tools in our mental toolbox to help us maintain the high standard of proficiency that the public deserves.

The Value of Case Studies

After the DC-10 crash landing in Sioux City, Iowa, in 1989, accident studies and flight tests conducted by McDonnell Douglas and NASA's Dryden Flight Research Center showed that engine thrust can be used as control in some cases and, thus, that it is extremely valuable for pilots to have practice with this technique so that they can use it during a crash landing, if necessary.

On November 22, 2003, an Airbus A300 struck by a ground-to-air missile was hit after takeoff from the Baghdad (Iraq) International Airport at an altitude of 8,000 ft. The flight crew notified the control tower, turned back, and made an emergency landing at the airport. This aircraft lost portions of its wing, including important control surfaces, and all of its hydraulic systems, including backup systems. Even worse, the aircraft's wing was on fire. The Airbus crew landed with no hydraulic systems, relying solely on engine power—just as the pilot of the DC-10 in Sioux City did when he reached that airport.

Aviation professionals familiar with the Baghdad incident have stated that part of the credit for the outstanding flying skills exhibited by the two Belgian pilots and British flight engineer was attributable to case studies: The Airbus captain recently had attended a safety seminar at which retired United Airlines Captain Al Haynes spoke. Haynes was the captain who managed, with his crew, to successfully land a damaged DC-10 with no hydraulic power (again, as in the Sioux City case study).

Remembering what he learned from Haynes's experience, the Airbus A300 captain and crew successfully landed their burning jumbo jet; after they landed, the fire was promptly extinguished by trained ARFF crews.

The Role of Case Studies in Preplanning

According to *Aviation Week and Space Technology* magazine (Dec. 1, p. 46), during the past 25 years there have been 35 shoulder-fired missile attacks on civil aircraft, twenty-four of which resulted in crashes with 500 fatalities. The U.S. Department of Homeland Security is choosing contractors to develop prototype missile–self-defense systems for use on commercial aircraft. These aircraft could employ laser, flare-and-chaff dispensers, or other self-defense systems (including new technology not publicly known at this writing). Case studies of such incidents can be invaluable in helping you preplan for aircraft crash rescues.

A useful learning tool is applying the information you've acquired from reading this book. Complete these exercises individually or as a group to further your learning process.

The goals of this section on case studies are to help you:

- Enhance and solidify the information you have obtained from this book.
- Realize that some aspects of an aircraft accident are absolute, for instance:
  - A deep crater at a crash site usually indicates little or no chance of survivors.
- Remember that water streams or Halon should never be directed on burning magnesium.
- Understand that more than one method can be the correct one to accomplish a task.
- Realize that many aircraft accidents have survivors.
- Remember, the outcome of a successful operation depends on planning ahead.

CASE STUDY 1: MID-AIR COLLISION AND CRASH

The Scene

This case study refers to the accident described in Chapter 1, involving two general aviation (private) airplanes that collided in mid-air over a remote area.

Aircraft 1 crashed into a grove of trees, with fatal results. No bystanders were available to respond to the crash except the farmer who owned the grove. There was no post-crash fire, and the debris was in large pieces.

Aircraft 2 was struck where the windshield meets the roof of the airplane, but it managed to land successfully on a remote dirt landing strip. The pilot suffered head injuries, yet was conscious and alert. The aircraft's control panel was broken loose from its location and forced backward, indicating that the pilot may have chest or other internal injuries. There was no post-crash fire, and many bystanders were available to help the pilot of this airplane because it took place near a busy highway.

Neither aircraft carried passengers.

Your Plan

1. Describe your actions if you were first in for aircraft 1.
2. Describe your actions if you were first in for aircraft 2.
3. Describe the nature of injuries you were likely to encounter and how you would respond.
4. How would you evaluate and treat suspected head and chest injuries?

CASE STUDY 2: CESSNA CRASH NOSE DOWN

The Scene

You have responded to a reported crash involving a Cessna 152 general aviation airplane. The terrain is desert sand. The aircraft is, effectively, standing on its nose, and its engine mount has broken. The engine is bent at a 90-degree angle. The stricken airplane has two passengers on board; both occupants are still strapped inside the aircraft.

Fuel is leaking from the aircraft and soaking into the sand, causing the airplane to begin to lean and become unstable. The two passengers are shouting to you that they are hurt, and they are becoming frantic to get out of the aircraft. The pilot complains he has pain in both legs and that his feet are "tangled in rudder pedals." There is no fire, because this was a low-impact crash.

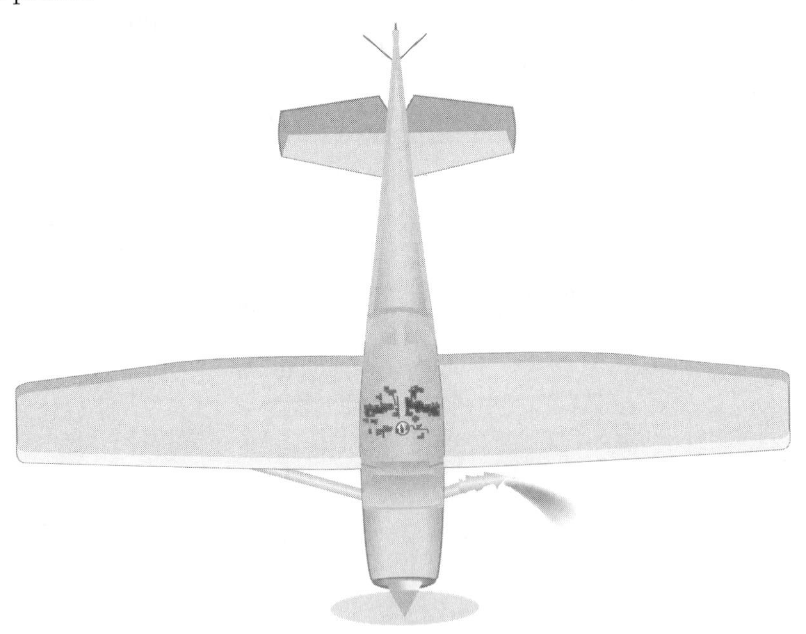

Your Plan

Refer to the illustration:

1. How would you stabilize the aircraft before extricating the occupants?
2. What are the potential risks to the airplane's stability if you apply foam or water to the spilled fuel?
3. Where would you place your firefighting vehicles and why?
4. What are the most likely injuries to the occupants?

CASE STUDY 3: MANAGING A COMPLEX SCENE

The Scene

In February 2005, a business jet with 10 passengers and crew failed in its takeoff attempt from the airport in Teterboro, New Jersey. The aircraft skidded across a busy six-lane highway, collided with two automobiles, and crashed through the wall of a warehouse. This resulted in multiple fires involving the airplane, automobiles, and the warehouse. The aircrew's quick thinking, the coordinated efforts of airport and non-airport fire agencies, and effective action by law enforcement personnel resulted in a successful and efficient firefighting and rescue operation. Although all eight passengers' clothing was soaked in jet fuel, the passengers and three crewmembers survived. Two motorists sustained injuries, as did several firefighters.

Based on the information in this book, this crash is an example of what severity of crash impact?

Your Plan

1. How would you deal with the following?
   - Passengers soaked with fuel
   - Traffic congestion caused by automobiles and pedestrian onlookers
2. Using your local emergency response plans for an aircraft incident, how would you interface with airport ARFF crews?
3. Who is your local authority to inspect damage to the building and allow business to return to normal?
4. Given the number of patients, how would you comply with your local EMS protocols?

CASE STUDY 4: DC-10 IN SIOUX CITY, IOWA

The Scene

In 1987, a disaster exercise was conducted as an airport mass casualty exercise. The scenario was a DC-10 crash, placed at the side of a runway and with 150 simulated survivors. (At that time, large-frame DC-10 airliners did not fly into this small city airport, but the recent unscheduled diversion of a DC-10 to Sioux City convinced local authorities that accidents involving such aircraft should be planned for.)

During the exercise, the director of the emergency services agency responding to the crash had identified several shortfalls and, as soon as practical, began to change the response plan to address these. These revisions included creating a mutual aid program by involving additional response agencies and other communities in the Sioux City's disaster plan. Representatives from all the participating agencies were encouraged to attend planning meetings. The disaster plan was updated on a recurring basis to address critical response concerns, resources, and other aspects of disaster scenarios.

Ironically, the 1987 exercise became real-world just two years after it had been conducted: On July 19, 1989, a DC-10 large-frame aircraft flying near Sioux City airspace lost all of its hydraulic systems, rendering its control surfaces useless. The crew somehow managed to steer the aircraft by adjusting the thrust settings on its two remaining engines and was able to crash land on the runway at Sioux City. The airplane literally tumbled down the runway as if a giant child had thrown it down in the midst of a temper tantrum. The broken aircraft finally came to a stop in a field of tall corn next to the runway.

When the stricken airplane contacted the ground, its cockpit tail broke off from the fuselage. Its right wingtip also broke off, causing fuel to spill and ignite as the remaining main portion of the aircraft spiraled down the runway. At first, the responders did not recognize the cockpit section, because it now resembled a huge, waist-high, crumbled beer can.

Survivors later recounted that not only were they exposed to flames and smoke, but also, as the airplane tumbled along during the crash landing, that their arms and legs were violently flailed about. Many of the survivors found themselves upside-down when the aircraft came to rest, because the fuselage was on lying on its back. Smoke and fire filled the cabin, and debris was scattered all over.

As one of the ambulatory survivors exited the mangled wreckage, he heard a baby crying, went back inside, and found her in an overhead luggage bin—the forces of the crash landing were so violent that she somehow wound up there. This brave man carried the baby out of the burning aircraft and returned her to her family, who had been forced to flee the airplane due to the fire and smoke. As in other airplane crashes, aircrew (in this case, flight attendants) and passengers were helping each other. Complicating escape and rescue, however, was that the fuselage was in a field of six feet tall corn, making it difficult to find both incapacitated and ambulatory passengers.

As a result of preplanning, training, and regular updating of response guidelines—and, critically—lessons learned from the previously conducted disaster drill, emergency rescue personnel carried out their duties smoothly and efficiently. By strange coincidence, the exact kind of aircraft impacted the runway, in the exact spot that was the simulated crash disaster.

The primary difference between the exercise and the real-world situation was the number of survivors: the training simulation included 150, but approximately 200 people survived the actual crash. Nonetheless, by having practiced the drill, the responders were able to effectively cope with the larger number of survivors.

As survivors began to escape the destroyed aircraft, the emergency medical services' well-organized mass casualty plan for an airplane crash went into effect. The local facilities in the trauma care system, including Marion Health Center and St. Luke's Hospital, did an outstanding job, in large part because they were staffed by double shifts and were able to improvise on the original plan, as needed.

The residents of this community were prepared and proactive, as well: More than 300 people showed up to donate blood even before a call for such help had been placed.

As soon as possible, United Airlines flew employees from other cities into Sioux City. These included passenger ticket and boarding agents, and reservations personnel, all of whom assisted survivors and their families in any way possible. By the morning after the crash, at least one United Airlines employee was present for every passenger's family, including the family members of deceased passengers.

In April 2004, Florida's AgSafe Web site contained a pertinent comment in the organization's newsletter, *Safety News and Notes* (vol. 5, no. 4). The article noted that the disaster planners and responders' tremendous response efficiency and success " . . . was only possible because of a drill that many had laughed at."

Your Plan

Review your community's emergency response plan and discuss how your agency would deal with the following issues:

1. How would you manage the following tactical concerns?
   - A large aircraft has broken into several large sections.
   - The flight deck (cockpit) is separated from the aircraft.
   - The crew is encased in tightly compacted, waist-high wreckage resembling a giant, crumpled beer can. You see an arm poking out of the wreckage and waving, as if to attract the attention of rescuers.

(*Continued*)

- The scene contains many fatalities and many survivors.
- A large portion of fuselage is resting in a field of tall corn. Many survivors have exited the cabin; some are lost among the corn. One manages to find his way out of the field and advises you that many survivors are wandering in the field, while some remain trapped in the portion of the airplane sitting in the field.

2. As a mutual aid responder to the airport, how would you make contact with incident command?
3. You are directed to resupply an ARFF vehicle by laying hose for a long distance, which will require tandem pumping. What should you consider as you drive your vehicle to the ARFF truck through a large debris area containing other rescuers, fatalities, injured and ambulatory survivors, and cargo and baggage?
4. Identify your nearest trauma centers and hospitals. How many patients can they accommodate?

CASE STUDY 5: FIRE IN A COMPOSITES AUTOCLAVE

The Scene

In the early 1990s, at a facility in northern California, a fire occurred in a large industrial autoclave used for the manufacturing aircraft parts made from carbon-fiber composites.

Your personnel successfully extinguish the fire in the autoclave using water fog.

Your Plan

Refer to the section in Chapter 3 on aircraft composite materials and discuss the following:

1. Why didn't the carbon-fiber layers delaminate and spread ash and particulate throughout the building?
2. What role did the adhesives used to glue the layers together play in this fire?

CASE STUDY 6: BURNED COMPOSITES

The Scene

In October 1990, a Royal Air Force Harrier combat jet crashed on a farm in Denmark. The local fire department extinguished the fire. The Royal Air Force dispatched a crash skin recovery team. The wreckage contained a considerable amount of burned carbon-fiber composite; it had been sprayed with diluted automobile "under seal" to contain loose fibers. In spite of these efforts, and even though responders wore protective clothing, masks, and goggles, many people in the rescue team suffered discomfort, sore eyes, and soreness in their throats and chests. After this incident, the RAF changed its procedures to require that a more stringent level of personnel protective equipment be worn during such incidents. A few months later, a similar accident involving the same aircraft type occurred in Germany. Because the more-stringent PPE requirements had been implemented, no one on the crash recovery teams was injured. These are only two examples of how the study of accidents and fires involving composites has resulted in better personal protective equipment and procedures.

Your Plan

1. What are some recognizable clues of a carbon-fiber mishap?
2. What simple method can fire departments use in order to temporarily contain loose ash and particulate until the first-in responders are relieved by hazardous materials or crash recovery crews?

CASE STUDY 7: THERE'S MORE OUT THERE THAN PEOPLE, FUEL, AND FIRE

The Scene

On a windy night, during a heavy thunderstorm, a jetliner crash landed, skidding off a runway and coming to rest near a river. The conditions made it difficult for ARFF crews to find the accident site: The crew was battling poor visibility, the increasingly intense storm, deep mud, and the rising waters of the river, which was overflowing its banks. As the firefighters attempted to extricate passengers and perform triage, they encountered poisonous snakes and alligators that were fleeing the flooding river. In addition, mutual aid units' response was delayed because of the inclement weather. Worst of all, perhaps, the airport firefighters' shiny, silver proximity suits attracted lightning.

Your Plan

1. If you are in a mutual aid unit responding to this incident, what is your airport plan for helping you navigate to this scene under such circumstances?
2. Are your vehicles equipped with forward-looking infrared radar (FLIR)?
3. Can your radios communicate with the ARFF vehicles?
4. Where do you fit into this plan if you are a member of one of the following organizations and you identify your point of contact (POC):
   - Private ambulance service
   - Non-airport law enforcement
   - Red Cross, Salvation Army, or a similar agency
   - News media
   - Chaplain or critical-incident stress management team manager
   - City or county government
5. Is there a predetermined staging point in such an incident

CASE STUDY 8: AIRLINER CRASH NEAR NEW ORLEANS

The Scene

This case study involves an incident that occurred in Kenner, Louisiana, during the hottest part of a summer day.

During takeoff, a large passenger jetliner crashed off the departure end of an airport runway and into a residential neighborhood.

The weather was *extremely* hot and humid, which adversely affected the operation because it placed the responders in danger of sustaining heat injuries. The crash site was densely populated and the neighborhood's streets were narrow. Despite the extensive devastation to nearby homes, from fires fed by large volumes of jet fuel, and massive loss of human life, the local and airport fire departments worked hard and prevented additional devastation within the impact area at this labor-intensive, challenging disaster.

Your Plan

Based on your occupation, outline how you would address the following concerns:

- Multiple house fires
- People trapped in damaged homes
- Traffic flow
- Incoming ARFF vehicles
- Access problems to houses and fire hydrants due to wreckage and pedestrians
- Large number of responders with heat-stress injuries
- Mass fatalities, which are excessively grim for responders and citizens to witness
- The influx of media reporters
- A long-term operation over many days
- The preservation of evidence in a situation where objects had to be moved to allow access to damaged homes

CASE STUDY 9: EXPLOSIVE ORDNANCE

The Scene

A military aircraft carrying a full load of explosive weapons became fully engulfed in flames for a brief amount of time. It was determined that there was enough time to promptly extinguish the blaze and ensure crew rescue. A few months later a very similar situation happened: The munitions exploded killing several firefighters.

Your Plan

1. By today's standards do we combat fires if munitions are engulfed in flames? Explain the reasoning for your answer.
2. What you would do?

CASE STUDY 10: MILITARY JET CRASHES INTO A HOTEL

The Scene

On October 20, 1987, an A-7D Corsair combat aircraft crashed into a hotel while attempting an emergency landing at the Indianapolis International Airport. The pilot successfully ejected when he realized his aircraft could not reach the airport runway. The aircraft cockpit and engine crashed through the hotel's main entrance, killing the staff at the lobby's registration desk. The aircraft's wings landed on a carport and in portions of the upper floor of the building. The aircraft was carrying approximately 20,000 lbs. of fuel, which quickly added to the magnitude of the resulting fire.

ARFF crews and equipment from the airport fire department arrived in approximately 1 minute and initiated a prompt and aggressive fire attack, as well as rescue operations.

Fire crews from the community (Wayne Township, Decatur, Indianapolis, and a local Air Force Reserve fire department) also arrived quickly. A unified command was established, and actions were well coordinated among all participating agencies.

Responders were greatly concerned about whether weapons may have been on board the aircraft. An Indianapolis police officer had to be dispatched to the hospital to obtain this information from the pilot; this delayed a comprehensive search for victims by almost an hour. Fortunately, there were no weapons on board the aircraft.

A National Defense Area was established. Personnel from the Air Force Reserve unit and local law enforcement worked together to ensure that a security perimeter was maintained around the aircraft wreckage.

The community had recently conducted several disaster exercises, and the area's disaster response agencies regularly trained with the airport disaster planners. Despite this effective preplanning, one shortcoming was identified: the need for more radio frequencies and for a common channel. An interim plan was devised, however, using runners and face-to-face communications at unified command.

Because the stricken airplane impacted the hotel registration desk, many of the hotel guests were unaccounted for. Command asked the local media to broadcast announcements asking any hotel guests to contact the local chapter of the American Red Cross and advise Red Cross personnel of their condition and whereabouts.

A county psychiatrist/psychologist was asked to conduct a debriefing session for all personnel affected by the disaster. Four departments participated in the incident. Another 20 agencies and business owners participated in supporting the fire-suppression, medical, and equipment/supply needs of this incident.

Your Plan

1. Discuss how your agency would handle an incident such as this one in your community.
2. Outline where and how you would obtain extra firefighting foam, if needed.
3. Determine whether you have enough aerial ladder apparatus (ladder trucks) for high-rise building rescue and firefighting.
4. A private ambulance company independently begins transporting patients to hospitals. How would you solve this problem?
5. Discuss your community's resources for dealing with the psychological aftermath of such a large-scale disaster, including PTSD.

TACTICAL EXERCISES

The following are a series of tactical exercises based on potential scenarios that you may encounter as an emergency responder. While the Case Studies in the preceding section are based on actual events and allow you to practice your skills of evaluating the incident and alternative solutions, these exercises require that you develop your own incident response plan based on the facts presented. Consider the facts carefully, as well as the main response components:

- Type of aircraft
- Terrain
- Potential/apparent hazards
- AHJ standards
- Local resources
- Unified command

SITUATION 1: GROUND EMERGENCY

Setup

A military fighter aircraft has made an emergency landing at a small airport.

- The incident is 500 feet from the nearest fire hydrant.
- The pilot has exited the aircraft and has been accounted for.
- The aircraft wheels are smoking; all that appears to be burning is fuel from a ruptured fuel line.
- Weapons have been discovered and determined to not be involved or burning.
- NTSB, military authorities, and support agencies on your checklist have been called by your dispatch and are en route.
- Firefighting engine 1 has a 50-gallon AFFF tank, as does E-2.

Your Plan

1. Draw an incident diagram and indicate the safest approach and positioning for equipment and people. Show the placement of all apparatus.
2. How would you resupply water to the "first-in" fire apparatus?
3. How would you supply water if the hydrant system is inoperable?
4. What other steps should you accomplish for this incident?

SITUATION 2: CRASH IN THE DIRT

Setup

A twin-engine passenger plane has crashed landed in a dirt field. Your objective is to keep fire away from the fuselage and protect passengers as they escape.

- The nearest water source is a fish hatchery pond 2 miles from your location.
- The nearest large-city fire department is 30 miles away.
- Most of the injuries are smoke inhalation and burns.
- One of the pilots is trapped inside the cockpit.
- You have made contact with a flight attendant.

Your Plan

1. Refer to the illustration: Indicate the danger areas by drawing in arrows and adding labels.
2. Where should pumpers and water tenders be positioned to assist the firefighters using water fog to control or extinguish the blaze?
3. How would you deal with the fuel flowing from the left wing?
4. List the agencies you would call on for assistance.
5. How would you account for all passengers who have left the aircraft?
6. How would you determine how many passengers remain inside the aircraft?

SITUATION 3: EMERGENCY LANDING

Setup

A twin-engine regional passenger jet with 37 people on board has made an emergency landing on a short runway at a small general aviation landing strip.

- There are no fire hydrants.
- The aircraft's tires are burning, and the fire is spreading to the aircraft's wings.
- Initial attack is accomplished using brush-fire apparatus.
- A water tender is resupplying the two pumpers that are on scene. Supply lines with clapper valves have been dropped in a straight (supply) lay to the incident, so additional apparatus can supply them with water.

- An Air Force base 3 miles away from the incident has dispatched an ARFF unit. Fortunately, this was a low-impact crash, and the fire has been controlled using the limited supply of AFFF in conjunction with high-pressure water fog properly applied. The result is that the fuse plugs have popped out of the aircraft rims, as designed, and your crews are limiting the fire area by directing hose streams to the aircraft wings and other points of flame impingement.
- All passengers have escaped.
- Several passengers have sprains and minor injuries, however, a 65-year-old passenger is complaining of extreme chest pains, is having difficulty breathing, and is semiconscious.

Your Plan

1. Refer to the illustration: Draw in arrows and label the safety concerns associated with parts of the airplane.
2. List the agencies you must notify and coordinate with.

SITUATION 4: A MILITARY CARGO JET CRASH

Setup

You are a captain with the local volunteer fire department, and you are driving home from work late at night. Your pager alerts you to a possible aircraft that has been downed in a wooded area 5 miles north of town. When you arrive on scene, you realize that a large-frame aircraft has crashed into a wooded area, and that streams and a small farmhouse are contained in the debris field.

- When you arrive, you see that the airplane is large, but the conditions at the scene prevent you from identifying the aircraft type.
- Your pager alerts you that the FAA and a nearby Air Force base have advised that a military cargo plane is off radar and missing. It is presumed down in the vicinity of the scene where you have arrived. You know from experience that the driving time to the nearby air base is 45 minutes.

Your Plan

1. What are your initial actions?
2. List the firefighting/rescue resources of your agency.
3. How does your mutual aid agreement address a situation such as this?
4. What should your strategic goals and tactical objectives be?
5. What specialized equipment or personnel would you request?
6. What outside agencies might you want to consider calling for help? What other concerns must be addressed before you terminate your agency's role in this incident?
7. Draw an incident site-safety plan including aircraft, buildings, and geographical features. Show the placement of equipment and resources within your community.

SITUATION 5: MID-AIR COLLISION

Setup

A large passenger jetliner collided with a military fighter jet 1.5 miles from your local airport.

- The jetliner has crashed in an industrial part of town, while the fighter jet crashed into a multistory building.
- There are mass fatalities, as well as survivors, from the aircraft and on the ground.
- The buildings around the crashed military aircraft must be evacuated, and a thorough search-and-rescue operation of the affected structures must be conducted.
- The military aircraft was not carrying munitions. The pilot ejected, and his parachute was found in a park by a nearby school.
- Looters at the scene are gathering suitcases and fleeing.
- Spilled jet fuel is running down the street into storm drains.

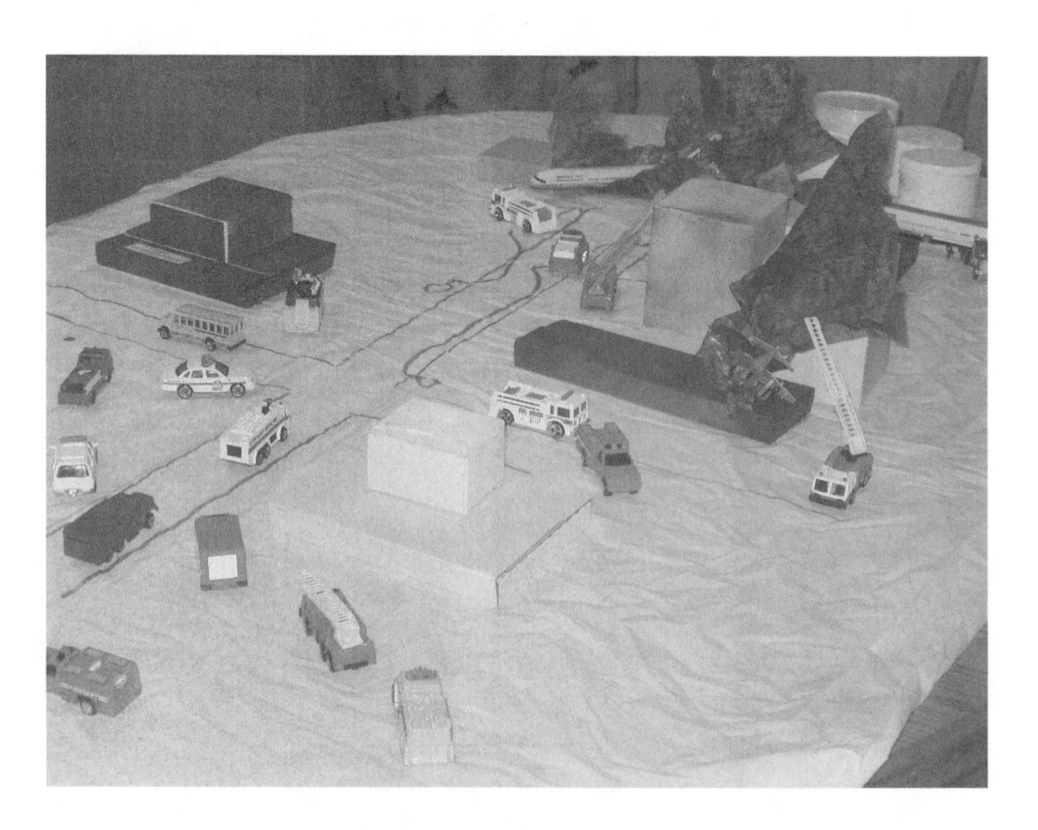

- Hysterical bystanders are hindering the operation.
- Citizen volunteers have offered assistance.
- Media reporters are asking questions of everyone.
- Military officials arrive.
- Civilian and military ARFF crews and apparatus arrive.
- FAA and NTSB officials arrive.

Your Plan

1. Explain how you would deal with these elements (within reason) as a member of one of the following groups:
   - Fire department
   - Law enforcement
   - Critical-incident stress management team
   - News media
   - Paramedic/EMT
   - Hospital administrator
   - Local disaster planner
   - Average citizen

SAMPLE CHECKLISTS AND REQUIREMENTS

Aircraft Accident Standard Operating Guidelines/Checklist

(May also be used for incidents involving aircraft composite materials.)
This is a generic checklist. Use as is or modify it to suit the needs of your agency.

- ❑ Assess the situation. Obtain aircraft type ASAP.
- ❑ Approach and position vehicles upwind and upgrade of smoke when possible.
- ❑ Direct follow-on responders to best approach (preferably upwind, upgrade).
- ❑ Isolate the incident and establish control at the incident site using an incident command system (ICS).
- ❑ Employ appropriate firefighting and rescue tactics based on the nature of this specific situation.
- ❑ Continue to reassess/update your size up throughout the incident.
- ❑ Monitor wind direction and speed.
- ❑ Maintain an activity incident journal.
- ❑ If possible, assign someone to take pictures of the scene when manpower permits, using the disposable cameras in your vehicles.
- ❑ Suggest downwind and lateral evacuation of smoke plume based on your judgment.
- ❑ Ensure that pressure vessels are kept cool.
- ❑ Note the presence of composites (if known): Take note of splintered debris, clumps of "hair," burned aircraft skin with visible cloth weave (carbon-fiber cluster).
- ❑ Record unusual fire behavior.
- ❑ Avoid directing high-pressure straight streams at burning composites if at all possible, to minimize breakup and spread of suspected composite-fiber particles. Use low-velocity straight streams (AFFF preferred) to penetrate layers of exposed composites where deep-seated smoldering fires may be located.
- ❑ Advise dispatch and other responders of the spread/direction of heavy smoke, or ash that you suspect contains composite particles.
- ❑ Request dispatch to make contact with FAA as soon as possible. They should ask the FAA for help to obtain:
    - ❑ Flight restrictions: No aircraft ground or flight operations within a minimum of 1,000 ft. radius and a minimum of 500 feet above the incident site. These distances may be increased based on your judgment.
    - ❑ Cargo contents (if known)
    - ❑ ETA of personnel from FAA, military (if applicable), and owning agency/person of involved aircraft.

(Continued)

- ❑ Gather as much specific information about the aircraft and the number of people on board.
- ❑ Request dispatch to confirm that the National Transportation Safety Board (NTSB) has been notified.
- ❑ As responding support agencies arrive on scene, have them sign the roster at incident command. Establish a unified command as soon as possible.
- ❑ Conduct thorough briefings to support agencies on a regular basis, before any new procedure begins or if the situation changes significantly.
- ❑ Use hazards analysis sketch (site safety plan) during briefings. Include a description of composite debris on the sketch.
- ❑ As soon as possible, instruct all non-firefighting vehicles to relocate to a safe area upwind (and upgrade if possible) from the incident.
- ❑ Minimize foot and vehicular traffic within the debris area to prevent the spread of composite ash or residue.
- ❑ Set up a hot zone (exclusion zone) based on your IDHA (identification and hazard assessment). Locate entry/exit control point a *minimum* of 25 feet from any suspected composite debris.
- ❑ Establish warm and cold zones using your local ICS or IMS system.
- ❑ Personnel within the fire/crash area must have full protective clothing and respiratory equipment, including self-contained breathing apparatus for all rescue personnel during any fire.
- ❑ Set up appropriate decontamination corridor. (No eating or drinking in the area.)
- ❑ Continue to be alert for any indications of burned composites, fibers, or ash that may be airborne, scattered on the ground, or in the proximity of the burned aircraft.
- ❑ Monitor downwind spread of any ash, heavy smoke concentration, or particulate.
- ❑ Check for deep-seated smoldering in large piles of burnt composite debris using a thermal imaging device, if available.
- ❑ Preserve evidence that may be of interest to incident investigation teams.
- ❑ Ensure that no debris is taken offsite.
- ❑ Direct witnesses to law enforcement officers to give initial statements and for coordination of contact with NTSB officials.
- ❑ Update data from appropriate agencies that may assist you in identifying: Specific aircraft type, hazardous cargo, advanced composite materials, or other significant data that may affect the safety of your personnel and the general public.
- ❑ Assess property involvement/damage/contamination caused by the fire and/or impact.
- ❑ If this incident is not an aircraft crash, establish contact with people from the building or facility involved.
- ❑ Ensure that runoff of firefighting agent is monitored and controlled. If debris from burned composites is suspected to be in this runoff, treat runoff as contaminated.

Post-Fire Checklist

This section is intended for use by disaster-response planners, environmental officials, firefighters, and anyone else who may be working at a post-crash/-fire scene.

It is understood that the majority of all fire departments do not perform hazardous materials cleanup; however, the incident commander is charged with ensuring that cleanup contractors are briefed on safety issues. The IC may be asked for input on correct disposal procedures.

- ❑ You can temporarily contain ash and loose particles of composites fibers with a fine water spray or a foam blanket, or by covering the material with sheets of plastic. Keep the crash/fire area damp with a foam blanket until other ash/particulate containment equipment is on scene. Polyacrylic floor wax may be sprayed on loose debris and particles; the best mixture is one part water to one part floor wax concentrate.
- ❑ If this incident happened in proximity of other aircraft:
  - ❑ Vacuum all air intakes with a HEPA vacuum cleaner.
  - ❑ Inspect all external access doors, vents, and hatches for signs of soot, particles, or ash. Vacuum as needed with a HEPA vacuum.
  - ❑ Closely inspect for smoke, ash, or particles that may have entered nearby parked aircraft. Both a visual and electronic ("sniffer"/LEL) checks of the interior should be conducted. If contamination is suspected, vacuum the interior thoroughly using a HEPA vacuum.
  - ❑ Before any exposed aircraft are permitted to fly; recommend that electrical checks and an engine run-up be performed.
- ❑ If buildings or other structures are exposed:
  - ❑ Thoroughly clean all antenna insulators, exposed transfer bushings, and so on.
  - ❑ Inspect air intakes in building or parked aircraft for soot deposits or other signs of contamination. Decontaminate these as needed.
- ❑ Personnel and equipment:
  - ❑ Decontaminate exposed personnel with HEPA vacuums, then wash PPE and turnout clothing.
  - ❑ Ensure that all involved personnel check themselves for possible puncture wounds and respiratory, eye, or skin irritation.
  - ❑ Exposed personnel should take cool showers.
  - ❑ Contaminated personal protective equipment (PPE) must be placed in sealed, airtight container(s).
  - ❑ Use HEPA-filter vacuums to remove loose ash or particulates from fire vehicles and other equipment.
  - ❑ Monitor decontamination water runoff.

(*Continued*)

- ❑ Transferring control of the site:
  - ❑ After the situation has been controlled, conduct a pass-on briefing with the authority assuming control of the site.
  - ❑ Confirm that the appropriate agencies have been thoroughly briefed on safety and protection of evidence.
  - ❑ Is a standby fire, rescue, or medic crew needed? Recovery operations should not begin until all firefighting and rescue operations are complete.
  - ❑ Terminate your agency's involvement when it is no longer needed and when the appropriate authority having jurisdiction has assumed command and control of the incident.
  - ❑ When command is transferred, supply the site safety plan (hazards sketch) during the safety/information debriefing.
- ❑ Additional guidelines:
  - ❑ All debris and contaminated materials must be disposed of in accordance with local, state, federal, and international guidelines. Coordinate these efforts with appropriate government agencies, as well as private environmental management officials. The Safety Investigation Board (SIB) and/or Accident Investigation Board (AIB) must advise when it is permitted to release the materials for disposal.
  - ❑ All contaminated materials must be in airtight containers and disposed of properly. They must be labeled with placards reading: "Composite Waste. Do Not Incinerate. Do Not Sell For Scrap," as shown in Figure B-1.
  - ❑ The appropriate agencies must ensure that any contamination is removed from soil. As incident commander, you must make other decisions, such as determining evacuation distances and the proper protective clothing for cleanup crews. If you have been relieved by military authorities at a military crash scene, this should be an easy transition, as they have well planned response checklists.

COMPOSITE DEBRIS
AIRCRAFT MSD __________
CONTENT __________
FIXANT TYPE __________
SPRAY PURPOSE __________

WARNING

COMBUSTIBLE SOLID. USE CLASS B EXTINGUISHMENT. DUST IS AND EYE, SKIN, AND INHALATION IRRITATION. FIBERS CAUSE PUNCTURE WOUNDS. BURNT DEBRIS HAS AN UNPLEASANT ODOR. DO NOT INCINERATE, DO NOT SELL FOR SCRAP. DISPOSE OF IN APPROVED LANDFILL.

CAUTION
CONTAINS COMPOSITE DUST AND FIBERS
AVOID CREATING DUST
DO NOT ENTER WORK AREA WITHOUT PROPER PROTECTION

Figure B-1 *Hazard markings for a composite-debris personal protective equipment (PPE) guide. (Courtesy of the U.S. Air Force.)*

Table 3.8-4 Personal Protective Equipment Considerations (1 of 3)

OPERATION	PROTECTION CONSIDERATION		EQUIPMENT
Clean up of randomly scattered small size composite debris: fragments, strips, clusters, pieces of sandwich structures, foam. Picking up small pieces scattered over a large area does not generate an appreciable amount of airborne particulate.	Eye and Face	Loose fibers protruding from the debris.	Goggles.
	Respiratory	—	—
	Head	—	—
	Foot	Puncture from sharp objects. Fluid spills. Harsh terrain.	Hard sole work shoes.
	Hand/ Forearm	Puncture wounds from protruding fibers. Residual dust, char material, or spilled fluids.	Inner long-cuffed nitrile glove. Outer leather glove.
	Body	Dirty and/or harsh environment.	Long sleeve BDU or protective coverall.

Figure B-2 *Personal protective equipment considerations (1 of 3). (Courtesy of the U.S. Air Force.)*

NOTE

Other operations occurring at the same time, site conditions, and weather may require additional protection.

Table 3.8-4 Personal Protective Equipment Considerations (2 of 3)

OPERATION	PROTECTION CONSIDERATION		EQUIPMENT
Backhoe, forklift or shoveling of damaged/burnt composite debris. Handling piles of damaged composites can produce appreciable amounts of airborne particulate. WITH OR WITHOUT FIXANT. Requirement is for direct contact with the debris. Requirement is for anyone within the workzone when piles of composite debris are being moved.	Eye and Face	Airborne composite dust and fibers of various sizes.	Goggles with half-face respirator.
	Respiratory	Airborne composite dust, fibrous particulate.	Full-face respirator preferred: combination filter (organic vapor and particulate 99% efficiency) for burnt debris and particulate for physically damaged debris.
	Head	Flying debris.	Hard hat.
	Foot	Puncture wounds from sharp objects, impact from carrying objects that can fall on the feet.	Hard sole and toe working shoes.
	Hand/ Forearm	Puncture wounds, residual dust, char material, or spilled fluids.	Inner-long-cuffed nitrile glove. Outer leather glove.
	Body	Generation of airborne composite dust, fibers, char material. Puncture wounds from fibers.	Protective coverall with hood, elastic wrist, and ankles. Booties are optional.

Figure B-3 *Personal protective equipment considerations (2 of 3). (Courtesy of the U.S. Air Force.)*

Table 3.8-4 Personal Protective Equipment Considerations (3 of 3)

OPERATION	PROTECTION CONSIDERATION		EQUIPMENT
Cutting the composite with a saw, pounding, or drilling. Cutting with a saw will generate heat, airborne particulate, and organic vapor, WITH OR WITHOUT FIXANT. Requirement is for the team in direct contact with the debris and anyone within close proximity.	Eye and Face	Airborne particulate	Goggles with half-face respirator.
	Respiratory	Airborne composite dust, fibrous particulates reaching the micron size range	Full-face respirator preferred: combination filter (organic vapor and particulate 99% efficiency) R series filter is needed when lubricants or cutting fluids are used.
	Head	Fibrous particulate	Protective coverall with hood
	Foot	Puncture from sharp objects	Hard sole and toe working shoes
	Hand/ Forearm	Puncture wounds, residual dust, char material, or spilled fluids.	Inner long-cuffed nitrile glove. Outer leather glove.
	Body	Generation of airborne particulate, fibers, char material. Puncture wounds from fibers.	Protective coverall with hood, elastic wrist, and ankles. Cutting with a saw may cause sparking. Sparks will destroy the protective coverall. An outer covering like rubber apron or leather chaps can be used to protect against sparks. Booties are optional.

Figure B-4 *Personal protective equipment considerations (3 of 3). (Courtesy of the U.S. Air Force.)*

Table 3.6-11 Summary of Garments™

EXPOSURE	GARMENT (not a comprehensive list)
Blood-borne pathogen	NexGen™ Tychem®
Chemical Liquid Hydrazine JP-8	 Tychem® -several types, Pro/shield 2, NexGen™ Tychem® F, SL and TK Tychem® TK (other Tychem® types were not tested)
Sharp object cut Fiber puncture	No garment listed provides 100% protection. Tychem® types are more rugged than Tyvek® Tyvek® types will provide adequate fiber protection for most fiber. Boron fiber may puncture Tyvek®
Dry particulate	Tyvek®, New Tyvek®, Tyvek® Type and 1431N and 1422 Tychem®, SL, QC Pro/shield 2, NexGen™
Radioactive dust	Tyvek®, Type 1422, Tychem® SL, New Tyvek®

Figure B-5 *A summary of personal protective equipment garments. (Courtesy of the U.S. Air Force.)*

Table 3.6-5

GLOVE TYPE	USE FOR
Neoprene	Freon Bumt Kapton®
Leather, Kevlar®	Boron fiber puncture Carbon fiber puncture Radioactive material
Nitrile	Composite dust JP-8 soot Bloodborne pathogens (BBP) Antifreeze Hydrazine Gasoline Kerosene Hydraulic fluid Jet fuel BBP decontamination solution
Thick nitrile	Carbon fiber
Latex	BBP
Butyl-rubber	Battery electrolyte Chemical warfare agents Antifreeze BBP decontamination solution

Lessons learned:

1. Never have enough.
2. Gloves too thin. Require frequent changes.
3. Use only powder-free gloves.
4. Get the right size. Mishap response involves manual labor. The wrong size will interfere with the job.
5. Close-fitting gloves should be used to avoid catching.
6. Short-cuffed gloves will not protect forearms from fiber puncture wounds.
7. Effectiveness of the plastic glove depends on the thickness.

Figure B-6
Suggestions for using personal protective equipment gloves. (Courtesy of the U.S. Air Force.)

Mishap response requires more than one type of glove. Leather, cowhide, or Kevlar is chosen for fiber puncture resistance and sharp or jagged metal and glass. Plastic gloves are chosen for chemical and particulate resistance and there are a number of plastic gloves to choose from. A glove selection guide has been tabulated, based on common mishap hazards, see above Table 3.6-5, Glove Types.

Table 3.7-6 Work Progress Survey

ACTIVITY	DEBRIS INFORMATION[1]	WORK CONDITIONS[2]	PPE
Physical Damage			
Fire			
Smoldering			
Overhaul			
Explosion			
Walking to the Site			
Visual inspection with hand movement			
Cutting, prying, sawing, pounding			
Search and Rescue			
Site Clean-up Picking pieces from topsoil Dissembling pieces Sorting Preparing for Transport Off-site Excavating Digging Sweeping			
Preparing for Transportation			
Transportation			
Storage			
Disposal Open storage boxes Sorting Cutting			

[1] Debris Information = type and amount of debris, condition of debris, location of debris.
[2] Work Conditions = Site hazards, weather conditions, terrain, type of work process and duration

Figure B-7 *A work progress survey (in checklist format). (Courtesy of the U.S. Air Force.)*

Table 3.7-4 Composite Assessment Questions

Composite Assessment Questions – Approximations are adequate			
Location	indoor outdoor		
Terrain	description		
Weather conditions at the time of occurrence	wind direction, velocity, temperature, description		
Cause of damage	Fire, physical, explosion		
Fire conditions: Where started Type/Aircraft location Size/Duration Ignition source Major fuel source Other burning materials Fire extinguishment Overhaul	 In-flight, impact, other Fireball, pool, smoldering		
Material Damage (if hazards sketch isn't available)	Description Site distribution or spread		
Work Process	Type		
Weather conditions at the time of work process	Wind direction, velocity temperature, description		
Air disturbances at the time of work process	Other than weather		
Weather Description Hail Sleet Fog Drizzle Rain Snow Thunderstorm Gusty wind Freezing rain Sunny	Composite form Terminology Tape Fabric Layer Stack Filament wound Solid laminate Sandwich laminate Core material Fibers Resin/matrix	Damaged Composite Terminology Fragment Strips Clusters Fiber bundles Dust Single fibers Scorched Delaminated	Work Process Overhaul Visual Inspection Cut, pry, saw, pound Search and rescue Cleanup Sorting Disposal Other

Figure B-8 *U.S. Air Force composite-fire assessment questions. (Courtesy of the U.S. Air Force.)*

GLOSSARY

Accident Investigators Investigators may work for a branch of the military or for the National Transportation Safety Board. They are tasked with piecing together all clues that may help determine the cause of the accident. Accident Investigators may be members of organizations such as the Airline Pilots Association and may also be FBI, ATF, CIA, or private-industry personnel, or contracted, self-employed consultants.

Aeromedical (medevac) operations Emergency aircraft (usually helicopter) operations that require teamwork among different agencies, including medevac aircrews, fire, police, ground-based EMS personnel, or military agencies. When helicopter crews are transferring patients, evacuees, or emergency officials, ground assistance is required to select a landing zone (LZ) that is safe and close to the personnel requiring transport.

Agency An organization or division of government with a specific jurisdiction or function offering a particular kind of assistance. In incident command structures (ICS), agencies are defined either as *jurisdictional* (having statutory responsibility for incident management) or as *assisting or cooperating* (providing resources or other assistance).

Agricultural aircraft These dispense products such as fertilizers, pesticides, herbicides, and seeds.

Aileron A control surface that consists of a moveable, hinged portion of an aircraft wing. Ailerons usually are part of the trailing edge of a wing. Their primary function is to help tilt the wings when an aircraft is banking.

Aircraft rescue firefighting (ARFF) Formerly called crash fire rescue (CFR).

Aircraft weight classes:

- **Heavy aircraft** can accommodate takeoff weights of more than 255,000 pounds; designed for long-distance flights.
- **Large aircraft** of more than 41,000 pounds, maximum certificated takeoff weight, up to 255,000 pounds.
- **Small aircraft** 41,000 pounds or less takeoff weight; carry small amounts of people and/or payload.

Airframe Generally, the parts of the aircraft having to do with the flight: the fuselage, boom, nacelles, cowlings, fairings, empennage, airfoil surfaces, landing gear, and so on.

Air traffic control The Federal Aviation Administration (FAA) division that operates control towers at major airports.

Airliner In the United States, an airliner is defined is an aircraft designated primarily for the transport of paying passengers. These aircraft are usually operated by an airline that operates under FAA regulations for scheduled or charter airline transportation.

Airport emergency plan exercise The Federal Aviation Administration (FAA) requires a full on-site aircraft disaster exercise once every three years (thus, this is often referred to as a "tri-annual"). These drills are used to evaluate the preparedness of an airport's emergency response teams and of surrounding municipal agencies in the event of a major disaster. Numerous local agencies are encouraged to participate, including ambulance services, EMS authorities, hospital trauma centers, the local American Red Cross, and officials from the Transportation Security Administration and the FAA.

Ambulatory A medical term that refers to patients who are able to walk.

APU See *auxiliary power unit.*

Aqueous film forming foam (AFFF) Fire-extinguishing agent that contains fluorocarbon surfactants and spreads a protective blanket of foam that extinguishes liquid hydrocarbon fuel fires by forming a self-sealing barrier between the fire and fire-sustaining oxygen. Note: Application of foam % means the percentage of concentrate in the final solution (e.g., 6% means 6% foam and 94% water).

ARFF Aircraft rescue firefighting (often pronounced "arf"). Formerly called *CFR,* which meant crash fire rescue.

Augmentee A person who volunteers, or is recruited, to assist in an emergency operation. These people can be utilized for nonhazardous duties, such as assisting with crowd control, obtaining ice, assisting in cutting wood for cribbing, and so on.

AVGAS High-octane gasoline with a flash point of –36° F.

Aviation Section This organization is a section of the NFPA. Its members are engaged in the design of aircraft and airport facilities, the operation of aircraft and airport facilities, and in protecting against and preventing loss or injury as applied to aircraft and airport facilities. In addition to defining codes and standards, this section is concerned with developing and promoting understanding of aircraft, airports, and other aviation sectors and their need for fire safety; as well as with promoting mutual understanding and cooperation related to aviation fire safety.

Auxiliary power unit (APU) This fuel-powered (usually turbine) unit supplies electrical power, air conditioning, and backup power to an aircraft during flight. The APU may also be used to power pneumatic (air) and hydraulic (fluid) pumps within the airplane.

Ballistic recovery system (BRS) This emergency device consists of explosive charges and hatch covers that, once activated, fire a rocket that drags a tightly compacted parachute up and toward the rear of the aircraft, enabling the airplane to float safely to the ground.

Black box This term refers to a box that contains the cockpit voice recorder or flight data recorder, which are important tools in an aircraft accident investigation. The black box is designed to withstand high-impact forces and is painted bright orange so as to make it highly visible. It is usually located in the tail of an airplane.

Body substance isolation (BSI) This term refers to wearing personal protective equipment (PPE), such as gloves, goggles, gowns, and masks, while rendering first aid or in any other way risking exposure to human body fluids.

Bogie A tandem arrangement of landing gear wheels. Bogies swivel up and down to enable all wheels to follow the ground as the altitude of the aircraft changes or as ground surface changes.

Boiling liquid expanding vapor explosion (BLEVE) Often called a "blevie," this dangerous event results when a closed container of liquid is exposed to excessive heat or flame impingement. The result is catastrophic container failure that often results in fragments of the container dispersing with violent force.

Bomb bay An enclosure in an aircraft fuselage whose doors open when bombs or other weapons are being released.

Bulkhead An upright partition that separates one aircraft compartment from another. Bulkheads may carry a part of the structural stress while forming the shape of an aircraft fuselage; equipment and accessories may be mounted on them.

Buddy store This aerial refueling system is used by the U.S. Navy S-3 Viking airplane. It uses a modified, externally mounted fuel tank with the drogue-parachute type of refueling system. It is attached to a weapons/bomb rack on aircraft and has been successfully tested by Boeing Aircraft, which used an A/F-18 Super Hornet jet to refuel other aircraft in flight.

Cabin Passenger compartment in an aircraft.

Canopy The transparent enclosure over the cockpit on some aircraft.

Cargo aircraft These aircraft, also called *transport aircraft,* are designated to carry freight and may be civil or military aircraft. Some cargo style aircraft have been converted for private, or VIP, operations.

CFR An obsolete term meaning "Crash Fire Rescue." Currently, in the United States, the acronym "CFR" refers to the Code of Federal Regulations.

Chaff This material deters missiles launched at aircraft from the ground or from other aircraft. During the Second World War, it was developed as a countermeasure, a method of confusing radar during combat. Aircraft (or other targets) disperse pieces of aluminum foil, metallized glass fiber, or plastic to form a cloud. Rescue personnel should preplan various aircraft to learn where chaff dispensers are located on them.

Civil aircraft This term refers to two categories of nonmilitary aviation: general aviation (private) and commercial. General aviation (GA) includes all civil flights, whether private or commercial.

Class A foam This firefighting foam is mixed with water, allowing the extinguishing agent to penetrate deep-seated fires involving matted grasses or other dry vegetation, bales of cotton, cardboard, or other Class A combustibles requiring a penetrating fire-extinguishing agent. It is also used to temporarily coat exposed flammable vegetation and structures, affording temporary protection from fire.

Class B foam This is used to control and extinguish fuel fires (Class B fires), such as those caused by gasoline or other liquid hydrocarbon fuels.

Cockpit Compartment where the pilots sit to fly the aircraft.

Cockpit shutdown procedures This term is often used to mean "emergency shutdown procedures," however, it is the methodical shutting off of aircraft engines, fuel pump switches, electrical power switches, hydraulic systems, and so on.

Cockpit voice recorder (CVR) This device records crew conversations and sounds in the flight deck (cockpit). It is intended for use in accident investigations, and it is contained within the aircraft's "black box".

Combat aircraft These military aircraft are designated for use in warfare and can carrying weapons systems. Examples include attack, bomber, electronic-warfare, and fighter aircraft.

Command post The location of the incident commander and command staff. It is at this location where those in charge of emergency crews and vital support personnel are briefed.

Commercial aircraft The official name for this category of aircraft is civil aircraft. Civil aircraft are most commonly used for transportation of people and freight for revenue on a scheduled or charter basis.

Commuter aircraft This term describes a small- or medium-frame aircraft that flies passengers on short routes.

Composite materials Refers to materials constructed by binding two or more unlike materials with resins, epoxies, glues, and so on. The result is strong yet lightweight material. Examples include, but are not limited to, carbon, fiberglass, and Kevlar composites.

Control surfaces Ailerons, flaps, elevator, rudder, and spoilers, which control an aircraft's direction of flight, altitude, and pitch.

Controls Any instruments or components provided to enable the pilot to control an aircraft's speed, direction of flight, altitude, power, and so on.

Critical incident stress management This short-term assistance focuses solely on an immediate and identifiable problem. Its goal is to enable affected employees to return to their normal daily routines as soon as possible, while decreasing their chances of developing PTSD.

Critical rescue firefighting and access area The primary response area for airport-based ARFF service.

Debriefing (psychological) A debriefing should take place within 72 hours of the incident. It gives individuals or a group the opportunity to talk about their experiences and how those experiences have affected them, and to discuss ways of dealing with their feelings about the incident. A debriefing usually is the second level of intervention for those directly affected by an incident, and often the first level of intervention for those not directly involved. During a debriefing, participants should be provided with information about the services available to them, and at-risk individuals should be identified. Post-debriefing follow-up should be done to ensure that everyone involved in an incident is feeling well or is referred to a professional counselor.

Defusings These are limited to individuals directly involved in the incident and are often done informally, sometimes at the scene. They are designed to quickly address responders' immediate needs.

Dry chemical A fire-extinguishing agent used to extinguish Class A, B, or C fires and composed of ammonium phosphate (ABC), potassium bicarbonate (brand name, Purple K), or similar material.

Dry powder A fire-extinguishing agent used for Class D (metal) fires and usually made of powdered graphite or similar materials. Brand names include Metyl-X™.

Ejection seat A seat with a rocket motor designed for quick escape from an aircraft.

Electrical system Usually located in the leading portion of the wing, electrical systems involve wiring in hard-to-access spaces. In a crash landing, electrical wires may be torn or damaged. Movement of the aircraft might cause a spark, which can ignite leaking fuel.

Elevator The movable horizontal portion of the tail of an aircraft. The elevator is hinged to the rear of the horizontal stabilizer and controlled by the pilot to move the aircraft's nose up or down or to level its flight position.

Emergency cut-in areas Designated locations, which enable rescuers to cut into an aircraft for removing entrapped personnel, or advancing hose lines.

Emergency escape route This applies to a safe escape path for passengers and crew of an aircraft crash landing to quickly escape with minimal chances of burns or sustaining secondary injuries as a result of escaping the aircraft. For firefighters or other emergency responders it is a designated escape route for them to exit a "danger zone" in the event it becomes too dangerous to continue their involvement in the operation.

Emergency evacuation The rapid exiting of an airplane during a situation posing threat of bodily harm or death.

Emergency locater transmitter (ELT) This device is designed to automatically send a radio signal from a stricken aircraft in the event of a crash. Rescue agencies such as the military and Civil Air Patrol can locate a downed aircraft by tracking the signal from an ELT.

Emergency power unit (EPU) Some fighter aircraft, such as the F-16, use an EPU instead of an auxiliary power unit. The EPU is powered by a toxic fuel called hydrazine, and, in the event of engine failure, it automatically starts and furnishes power for flight instruments and aircraft control movements.

Emergency response plan (ERP) (Also, local emergency response plan [LERP]) The plan maintained by various jurisdictional levels for managing a wide variety of potential hazards.

Emergency response provider Includes federal, state, local, and tribal emergency public safety, law enforcement, emergency response, emergency medical (including hospital emergency facilities), and related personnel, agencies, and authorities. Also called an *emergency responder.*

Emergency shutdown procedures These are methodical and often sequential steps for shutting down an airplane. They include shutting the aircraft throttle (or throttles) to the idle, then "off" position; turning off electrical systems; safetying ejection seats (if possible); turning off fuel selection switches, and so on.

Empennage An aeronautical term referring to the complete tail assembly of an aircraft and its parts or components, including the horizontal stabilizer, elevators, rudder, and so on.

Employee assistance program (EAP) A program offered by an employer to help employees address personal problems that might negatively affect their job performance or their physical and mental health. EAPs vary, but, ideally, they should address substance abuse, emotional distress, exaggerated safety concerns, physical health, dysfunction affecting marriage and family, death and grief, financial problems, legal problems, and other life-affecting events. These services usually are prepaid by employers, often through a third-party provider, and free to employees and their household members.

All information obtained during or generated by these programs is maintained in confidentiality. Employers usually are not informed of which employees are using EAP programs, unless there are extenuating circumstances that require an employer to be given this information. In that case, the proper release forms must be signed by the employee. In some cases, management may require employees to seek EAP assistance because their behavior or work performance has become unacceptable.

Engine The motive that power the airplane and allows it to travel. May be piston-driven propeller, turboprop, or jet engines.

Evacuation Organized, phased, and supervised withdrawal, dispersal, or removal of civilians from dangerous or potentially dangerous areas, and their reception and care in safe areas.

Face-to-face briefing Refers to an in-person, verbal briefing from one person to another.

Federal Aviation Administration (FAA) This branch of the Department of Transportation (DOT) promotes aviation safety by establishing safety recommendations, rules, and regulations that involve aircraft and the aviation industry.

Federal Emergency Management Agency (FEMA) Now part of the Department of Homeland Security, FEMA is intended to create a more efficient system to help America prepare for, respond to, and recover from all forms disasters.

Field notes These handwritten notes include multiple observations of an aircraft incident, including the time, rough sketches of initial findings, and significant events. During the initial phases of setting up command, these notes may be written in a simple notebook or paper, but they should later be compiled in an incident journal.

Fire-extinguishing agent The material used to extinguish a fire, such as Class B foam, Class D dry powder, water fog, hose streams, gaseous agents (e.g., carbon dioxide or Halon), or a combined dry chemical-water application.

Firewall A bulkhead separating two compartments of an aircraft, for example, the engine compartment and the aircraft's cockpit/cabin.

First responder Local and nongovernmental police, fire, and emergency personnel who are responsible during the early stages of an incident for the protection and preservation of life, property, evidence, and the environment, including emergency response providers as described in the Homeland Security Act of 2002, as well as emergency management, public health, clinical care, public works, and other skilled support personnel (such as equipment operators) who provide immediate support services during prevention, response, and recovery operations. First responders may include personnel from federal, state, local, tribal, or nongovernmental organizations.

Fixed-base operator (FBO) An entity at an airport, which renders maintenance, storage, and servicing of aircraft. They may also rent aircraft to licensed pilots.

Fixed-wing aircraft Airplanes that consist of a fuselage, wings, and a tail assembly.

Flaps These adjustable airfoils are attached to the leading or trailing edge of the wings, affecting the aircraft's aerodynamic performance during landings and takeoffs. Usually extended during landings and takeoffs.

Flight data recorder (FDR) This device records in-flight information such as speed, engine RPM, aircraft flight attitude, pitch, and so on. It also can record outside air temperature, vertical acceleration, and other variables while an aircraft is in the air. It may also be referred to as a *flight recorder* or *digital flight data recorder*.

Flight deck The pilot's compartment of a large airplane. Also referred to as a *cockpit*.

Fuel Hydrocarbons that power aircraft engines. May be kerosene-based jet fuels or aviation gasoline.

Fuselage This is the main body of the aircraft, to which the wings and tail are attached. Usually, construction incorporates frames and bulkheads that make up the individual compartments or sections of an airplane and strengthen the fuselage.

General aviation (GA) Refers to all air activities and aircraft that are not associated with scheduled commercial aircraft operations or military aviation.

General aviation aircraft All civil aviation aircraft used for private, unscheduled, non-revenue operations. The most common examples are Cessna, Piper, and Beech. This category also includes lighter-than-air balloons, air ships, sail planes, gliders, rotorcraft (helicopters), and gyroplanes. The majority of all aircraft flights in the United States involve general aviation aircraft.

Hatches Openings that provide a means for escape from, and emergency entry into, a distressed aircraft. They may be located at the sides, bottom, or top of the fuselage. Hatches are usually built into larger general aviation aircraft and commercial and military aircraft that carry passengers or cargo. Hatches generally have controls allowing them to be operated from inside or outside of an airplane by using quick-opening compression devices.

Hazard Something that is potentially dangerous or harmful, or that may result in an unwanted outcome.

Hazardous material Any substance or material that has been determined by the Secretary of Transportation as capable of posing an unreasonable risk to health, safety, and property when transported in commerce.

HE (pronounced "H-E") An abbreviated term for high explosives, which are contained in ammunition, bombs, and missiles or cannon shells.

High-impact crash During this level of crash, aircraft strike the ground with tremendous and violent force. These crashes are characterized by small pieces of debris, with little or no large wreckage. Chances of survivors are almost nonexistent.

Hydraulic system The tubes, hoses, lines, and pumps through which hydraulic fluids or oils pass to operate control surfaces, landing gear, and so on.

Hydrazine A fuel used in the emergency power unit on aircraft such as the F-16. This substance is extremely toxic and caustic.

Idle The lowest possible speed at which an engine operates or runs.

Incident An occurrence or event, natural or human-caused, that requires an emergency response to protect life or property. Incidents can include major disasters, emergencies, terrorist attacks, terrorist threats, wildland and urban fires, floods, hazardous materials spills, nuclear accidents, aircraft accidents, earthquakes, hurricanes, tornadoes, tropical storms, war-related disasters, public health and medical emergencies, and other occurrences requiring an emergency response.

Incident command This determines the success or failure of an emergency response. Incident command is especially critical during aircraft accident rescue, because crew and passengers are encased in these craft and surrounded by large volumes of flammable aviation fuel, as well as baggage and cargo that may contain hazardous materials.

Incident commander (IC) The person who is designated as being responsible for all activities associated with an incident, including the development of tactical plans, strategies, and the utilization of resources. The IC has overall authority and responsibility for managing incident operations and is responsible for the direction of all incident operations at the incident site. He/she also has the authority to demobilize response personnel, equipment and other resources as the situation reduces in size, magnitude, or degree of hazard.

Incident command post (ICP) The field location at which the primary tactical-level, on-scene incident command functions are performed. The ICP may be colocated with the incident base or other incident facilities and is *normally* identified by a green rotating or flashing light.

Incident command system (ICS) A standardized on-scene emergency management system designed to provide an integrated organizational structure that reflects the complexity and demands of single or multiple incidents, without being hindered by jurisdictional boundaries. ICS is the combination of facilities, equipment, personnel, procedures, and communications operating with a common organizational structure to aid in the management of resources during incidents. ICS is used for all kinds of emergencies, and it is designed to be in compliance with NIMS.

Incident diagram This is a sketch of the scene of an emergency incident, including aircraft, rail, highway, pipeline, maritime, HAZMAT, wildfire, structural fire, or other accidents or incidents.

Incident site safety plan A site safety plan is a drawing or diagram of an incident, such as a HAZMAT, accident, or other dangerous incident. The plan must comply with safety procedures and includes exclusion zones, division assignments, emergency escape routes, a decontamination corridor, and safe-refuge areas. Before an entry is made into a HAZMAT, a safety briefing is conducted and all participants are advised of the action plan and of emergency safety contingencies specific to the incident area.

In-flight emergency In this event, the aircraft pilot declares an emergency situation while the aircraft is flying. The emergency may include hydraulic failure, smoke smell, engine fire, and so on. The nature of this emergency has an adverse effect on the safe operation of an aircraft, therefore, rescue vehicles are positioned at predetermined strategic locations at the airport runway and placed on standby alert. Referred to as an *Alert II.*

Initial response Resources initially dispatched and committed to an incident.

Intake area The potentially dangerous area in front of and at the sides of a jet engine where the engine sucks in air with tremendous force.

International Air Transport Association (IATA) This international industry trade group of airlines is based, with ICAO, in Montreal, Quebec, Canada. Today, IATA has more than 270 members from more than 140 nations. This organization helps airline companies in such areas as pricing uniformity and establishing regulations for the shipping of dangerous goods, and it publishes the all-important *IATA Dangerous Goods Regulations* manual. This manual is recognized worldwide as the field source reference for airlines shipping hazardous materials referred to in aviation as dangerous goods.

International Civil Aviation Organization (ICAO) A specialized agency of the United Nations that is responsible for developing international rules governing all areas of civil aviation. Rules relating to transportation of commodities must comply with Title 49 CFR. The ICAO's safety responsibilities include a regulatory framework, enforcement and inspection procedures, and, when necessary, corrective measures associated with airworthiness of aircraft, airport safety, personnel licensing, and international aviation rules.

Jet A fuel Kerosene-type fuel with a flash point of 110° F to 115° F.

Jet B fuel Kerosene-type fuel with a flash point of –16° F to 30° F.

Jet engine intake The portion of a jet engine where air enters in at great velocity. The engine intake poses a potential hazard for responders.

Jetliner This term commonly describes jet-powered cargo or passenger-carrying aircraft.

Jettison To "blow away, or discard, from the aircraft certain items such as canopies, external fuel tanks, or externally mounted weapons/ordnance.

JP-4 fuel 65% gas and 35% light petroleum distillate, flash point –10 F to 30 F.

JP-5 fuel Specially refined kerosene, flash point 95 F to 145 F.

JP-8 fuel Has the same properties and burn characteristics as Jet-A fuel.

Jurisdiction A geographic area or realm of authority; it may be based on legal responsibilities and authorities. Jurisdictional authority at an incident can be political or geographic (e.g., city, county, tribal, state, or federal boundary lines), or functional (e.g., law enforcement, NTSB, military, public health).

Landing gear The understructure—usually, wheels, tires, and struts—supporting the weight of an aircraft when it is not in the sky. Also called the *undercarriage.*

Landing zone (LZ) An area designated for the transfer of medical patients or evacuees from a ground area to board an aircraft, usually a helicopter. The landing zone may be a roadway, school, parking lot, or open field. The surrounding area should be clear of street lamps, trees, power poles, buildings, fences, wires, radio towers, or other obstacles.

Large-frame aircraft Very big airplanes such as Boeing 747, 767, 777, 787, and MD-11; Airbus A300, A330, A340, A-350, and A-380; Antonov 124, C-5, C-141, B-1, B-2, and B-52.

Leading/trailing edge The forward/rear edge of an airfoil. Applies to tail surfaces, wings, propeller blades, and so on.

Level of impact This refers to the speed of the aircraft when an aircraft struck the ground:

High-impact crash Characterized by extensive break-up and disintegration of the airplane. Most of the wreckage will be in small pieces. It is likely there will be a crater at the impact site. It is almost impossible for an aircraft occupant to survive a high impact crash.

Medium-impact crash Characterized by the aircraft fuselage breaking apart into several large pieces. The likelihood of a post-crash fire is great. There is a good chance of survivors, but they are likely to have sustained moderate to severe trauma.

Low-impact crash Can be either on or off and an airport runway such as a county road or open field. The fuselage is basically intact, and it is common for people to survive this type of crash. Most likely you will encounter injured people and fires on the ground as well as inside the aircraft.

Life safety This term refers to the protection of human life.

Local government A county, a municipality, a city, a town, a township, a local public authority, a school district, a special district, an intrastate district, a council of governments (regardless of whether the council of governments is incorporated as a nonprofit corporation under state law), a regional or interstate government entity, or an agency or instrumentality of a local government; an Indian tribe or authorized tribal organization; or a rural community, an unincorporated town or village, or other public entity.

Longerons The principal longitudinal (lengthwise) structural members of the fuselage.

Liquid oxygen (LOX) Oxygen that has undergone a cryogenic process that freezes the gas, thereby compressing it so that it can be stored in a smaller tank or reservoir. LOX is extremely cold and hazardous.

Magneto This device is found in aircraft piston-type aircraft engines and generates electric current from a magnet that spins as the motor crankshaft operates. If the aircraft's electrical system fails, the magnetos ensure that the engine spark plugs get electricity. Engines usually contain two sets of magnetos.

Mass casualty incident (MCI) Any accident or catastrophe that involves large numbers of casualties.

Medium-frame aircraft Examples of these aircraft are the ATR72, CRJ-700, DC-9, and C-130. These are used for passenger or cargo transport, or specialized training.

Microjets This term refers to a class of aircraft called very light jets (VLJs).

Military aircraft These may be any kind of airplane owned and operated by military forces.

Mitigation This term describes activities designed to reduce or eliminate risks to persons or property or to lessen the actual or potential effects or consequences of an incident. Mitigation measures may be implemented prior to, during, or after an incident. Mitigation involves ongoing actions to reduce exposure to, probability of, or potential loss from hazards. Fire suppression, rescue, and managing hazardous materials release may fall under this descriptive term.

Mutual aid Assistance rendered by one agency to another agency.

National Defense Area (NDA) The NDA is an area established on nonfederal lands located within the United States or its possessions or territories. The purpose is for safeguarding classified defense information or protecting Department of Defense (DOD) equipment and/or material by keeping non-authorized people away from such sensitive materials. It becomes temporarily in control of the DOD.

National Fire Protection Association (NFPA) This is an independent, voluntary, and nonprofit organization whose goal is to reduce the loss of lives and property resulting from aircraft emergencies. It is a source of research and education for all subject areas as they relate to fire and its prevention. The association develops codes and recommendations for fire safety standards. It is composed of various committees, such as the Aviation Section (which is of particular interest in this book).

National Incident Management System (NIMS) In compliance with federal law, the NIMS establishes standardized incident management processes, protocols, and procedures that are applicable for all responders (federal, state, tribal, and local).

National Institute of Safety and Health (NIOSH) Part of the U.S. Department of Health and Human Services, NIOSH operates within the Centers for Disease Control and Prevention (CDC), and is the federal agency responsible for conducting research and making recommendations for the prevention of work-related injury and illness.

The *Occupational Safety and Health Act (OSHA)*, established in 1970, created NIOSH and OSHA. Different states, however, take different approaches to legislation, regulation, and enforcement of these regulations.

Similar worker-safety provisions also are enforced in Canada and in many European Union countries, which have enforcing authorities to ensure that the basic legal requirements relating to occupational safety and health are met ensuring good OSH performance.

In Canada, the Canadian Centre for Occupational Health and Safety (CCOHS) was created based on the belief that all Canadians had "... a fundamental right to a healthy and safe working environment." CCOHS is mandated to promote safe and healthy workplaces to help prevent work-related injuries and illnesses.

The European Agency for Safety and Health at Work (EASHW) was founded in 1996. In the UK, health and safety legislation is drawn up and enforced by the Health and Safety Executive under the Health and Safety at Work Act of 1974.

National Transportation Safety Board (NTSB) The federal agency charged with investigating and determining the reason for and contributing factors to aircraft, railroad, maritime, highway, and pipeline transportation accidents. This agency makes recommendations to facilitate the prevention of additional mishaps or accidents.

Nongovernmental organization (NGO) A nonprofit entity focused on the interests of its members, individuals, or institutions so as to serve a public purpose, not benefit a private entity. NGOs are not created by a government, but may work cooperatively with governments. Examples of NGOs include faith-based charity organizations and the American Red Cross.

Occupants Passengers and aircrew on board an aircraft. Also referred to as "souls on board" (SOB).

Occupational exposure This term is associated with universal precautions and refers to unprotected contact with potentially infectious materials that may occur during the performance of an employee's duties.

Ordnance This term refers to ammunition, bombs, rockets, or other explosive materials that may be carried on military aircraft.

Personal protective equipment Equipment intended to protect rescue personnel from injuries. It includes head gear, special boots or shoes, clothing, and respiratory masks.

Phonetic alphabet A spoken-word alphabet designed to eliminate confusion for listeners when certain letters sound alike.

A – Alpha	J – Juliet	S – Sierra
B – Bravo	K – Kilo	T – Tango
C – Charlie	L – Lima	U – Umbrella
D – Delta	M – Mike	V – Victor
E – Echo	N – November	W – Whiskey
F – Foxtrot	O – Oscar	X – X-ray
G – Golf	P – Papa	Y – Yankee
H – Hotel	Q – Quebec	Z – Zebra
I – India	R – Romeo	

Pitot tube A hollow, protruding tube that resembles a gun barrel and is attached to a wing or the nose of an aircraft. It may be shaped like the letter "L." It measures RAM air pressure and translates those measurements into air speed. (This is discussed in detail later.) The tube, which is equipped with an internal heating device to prevent freezing, conveys the air speed information to display instruments on the control panel.

Pooled-fuel fire This term refers to a fire in a pool of spilled liquid fuel.

Post-incident critique This session examines which parts of an operation went well and according to plan and which did not: It is used to identify operational shortcomings in a constructive manner and to design methods for overcoming those shortcomings. Many agencies prefer to conduct post-incident critiques in a debriefing format. This focuses on addressing tactical/operational events and is different than a debriefing that is intended to address the fire suppression/rescue-operation from a tactical standpoint.

Post-traumatic stress disorder (PTSD) This term refers to a specific set of emotional and psychological distress symptoms resulting from involvement with a stressful or traumatic event. For first responders, the most stressful events commonly are line-of-duty deaths, coworker suicide, multiple-casualty incidents, and large-scale disasters. Responders who have experienced a traumatic event may develop PTSD if intervention by trained counselors is delayed.

Symptoms of PTSD may involve physical or psychological problems such as insomnia, sudden changes in weight (loss or gain), anxiety, depression, uncontrollable anger, withdrawal, lack of any emotion, nightmares and flashbacks, or hypervigilance and exaggerated "startle response."

Most people who experience traumatic events do not develop PTSD, which is primarily an anxiety disorder and should not be confused with the normal grief and adjustment that occur after traumatic events. For additional information, contact your local mental health association.

Preplanning The identification of hazards and appropriate mitigating actions that may be associated with an emergency event, such as an aircraft crash, a hazardous materials incident, a building fire, or a natural disaster. Preplanning commonly defines roles, resources, and individual responsibilities of emergency response personnel, aircrew, or key personnel in buildings or other facilities.

Private sector Organizations and entities that are not part of any governmental structure. These include for-profit and not-for-profit organizations, formal and informal structures, commerce and industry, private emergency response organizations, and private voluntary organizations.

Propeller A rotating airfoil, or "screw," attached to a hub, which when rotated, propels air, thus enabling the aircraft to move.

Propwash (slipstream) Wind driven astern by an operating propeller.

Public information officer During an emergency, the incident commander designates a public information officer (PIO) to ensure that information provided to the public by any means available, usually by the media. Response organizations coordinate information with PIOs and clear all press releases with incident command before releasing information. PIOs function as the official point of contact for the media, and keep them advised of proper and accurate information related to all phases of an incident. This information should be in concert with the incident commander and members of the unified command staff. When responders are approached by the media, they should refer them to the designated PIO.

Raindrop effect The most efficient aqueous film forming foam (AFFF) firefighting technique. The raindrops do not form when leaving the nozzle or turret, but when they contact the burning fuel. The stream must be adjusted so it is not so sparse that it evaporates or is carried away by heat updrafts from the fire.

Regional airliners These airliners are intended to carry between 35 and 100 passengers. Although these aircraft are used by commuter and other small airline companies, many of them also are in service with major airlines. The term *regional jet* refers to a jet-powered airliner in this category.

Resources Personnel, equipment, supplies, and facilities available or potentially available for assignment to incident operations and for which status is maintained. Resources are described by kind and type and may be used in operational support or supervisory capacities at an incident or at an EOC.

Response Activities that address the short-term, direct effects of an incident. Response includes immediate actions to save lives, protect property, and meet basic human needs, as well as the execution of emergency operations plans and of incident mitigation activities designed to limit the loss of life, personal injury, property damage, and other unfavorable outcomes. As indicated by the situation, response activities include applying intelligence and other information to lessen the effects or consequences of an incident; increased security operations; continuing investigations into the nature and source of the threat, ongoing public health and agricultural surveillance testing processes; immunizations, isolation, or quarantine; and specific law enforcement operation aimed at preempting, interdicting, or disrupting illegal activity, and apprehending actual perpetrators and bringing them to justice.

Retractable landing gear Landing gear that may be withdrawn into the body or wings of an aircraft to reduce drag while the aircraft is in flight.

Ribs A part of the skeletal structure of an aircraft wing that gives the airplane form, strength, and shape.

Rotor Rotating airfoil assembly (propeller) of helicopters.

Rudder The upright, movable part of the tail assembly, which controls the direction of the aircraft.

Short take off and landing (STOL) This class of aircraft is designed to take off or land on short runways.

Simple task This is a task that removes a person from a danger zone and relocates them in a safer area. This task may be directing traffic, serving meals to tired rescuers, or a similar assignment.

Skin The outer covering of an aircraft, including the fuselage and wings.

Slats Movable auxiliary airfoils whose primary function is to increase the aircraft's stability. This term usually references the fixed horizontal tail surface of the aircraft.

Small-frame aircraft These carry small numbers of people and/or amounts of payload. Examples would be a Cessna 172, Lear 21, A-37, T-37, and Mooney 201.

Spars The principal structural members, or beams, of a wing.

Specialized aircraft These are used for duties not found in mainstream commercial or military aviation, such as weather data gathering, drug interdiction, agricultural spraying and seed planting, law enforcement, medical evacuation, aerial firefighting, and other tasks.

Stabilizers These components are like fins, and they house the rudder that controls back-and forth movements (called *yaw*).

Standard Operating Guideline (SOG) A SOG provides guidance and suggested procedures for managing emergency responses or guidelines for organizational operating practices. It defers to a person's experience, common sense, and good judgment and provides greater flexibility in making tactical decisions than does a Standard Operating Procedure.

Standard Operating Procedure (SOP) A SOP is more binding than a Standard Operating Guideline (SOG). SOPs outline explicit procedures and policies that must be followed during an emergency response or other job related duties, and regulations.

Stringer A long, heavy horizontal timber used for any of several connective or supportive purposes.

Strut Structural members that are designed to resist pressure in the direction of their length (e.g., landing gear struts).

Surface wind A common cause of helicopter accidents involves inaccurate or unavailable information on surface wind direction and speed at landing sites. Aircraft land and take off into the wind, so the pilot must know the direction (and speed, if possible) of the prevailing surface winds.

Tactics Tactics is the "doing" of all the necessary operational activities needed to accomplish the goals set by the IC. Tactics must follow the goals and objectives set forth by the strategic plan.

Tail This portion of an aircraft consists of vertical and horizontal stabilizers, rudders, and elevators.

Terrorism Any act that endangers human life or is potentially destructive to critical community infrastructure or key resources, in violation of the criminal laws of any nation with jurisdiction where such an act occurs. These acts are intended to intimidate or coerce a civilian population; to influence the policy of a government by intimidation or coercion; or to affect the conduct of a government by mass destruction, assassination, or kidnapping.

Thermal runaway condition This is an electrochemical reaction that causes a battery to overheat, release toxic vapors, spew electrolyte, and very likely explode. An aircraft suffering this problem must land as soon as possible.

Threat An indication of possible violence, harm, or danger.

Tilt-rotor aircraft Called "powered-lift vehicles," these vehicles can take off, land, and hover similarly to a helicopter, and then can change wing angle and fly like a turboprop airplane.

Trailing edge The rear edge of an airfoil. Applies to tail surfaces, wings, propeller blades, and so on.

Triage This term means "to sort out." It is used in emergency medicine, especially during mass-casualty incidents and in crowded hospital emergency rooms. It is the process of sorting patients for priority treatment based on factors such as severity of injury, likelihood of survival, and available resources. Systems of designation include ranging systems (from "good to grave" or "1 to 5") and color codes: Use whichever system is common to your area. The START (simple triage and rapid treatment) system is very popular and is taught by the American Red Cross and EMS teaching institutions.

Transport aircraft This term refers to aircraft designated for the purpose of carrying passengers or cargo (freight).

Turbine engine An engine that does not have pistons. The several types of turbine engines include:

Turboprop—The propeller shaft is turned by a turbine shaft.

Turbojet—An arrangement of compressor blades that are turned by shafts propelled by high-speed gases.

Turbofan—Similar to the older turbojet engine, a turbofan contains a large fan located at the front of the engine, which increases airflow into the engine and generates greater thrust.

Turboshaft—These engines power helicopters.

Turbine engines may be used to power aircraft auxiliary power units or ground power equipment, such as electrical power generators.

Ultralight aircraft This class of aircraft is characterized by a small airframe and motor. These machines are designed for recreational purposes and do not require an FAA registration number, airworthiness certificate, or pilot certification. These aircraft resemble hang gliders, and most have a single-seat configuration. Some ultralight aircraft, however, are designed for a pilot and one passenger.

Unified command structure An application of ICS used when there is more than one agency with incident jurisdiction at a scene, or during an incident with cross-political jurisdictions. Each agency designates an incident commander at a single *incident command post (ICP).* The agencies work through these designated members of the unified command to establish a common set of objectives and strategies and a single *incident action plan (IAP).*

Universal precautions These work practices comply with OSHA regulations and includes wearing body substance isolation equipment and clothing. Universal precautions include engineering controls (built-in protection), wearing personal protective equipment (PPE), decontamination, and properly disposing contaminated materials.

Vertical stabilizer This component is like a fin, and it houses the rudder that controls back-and-forth movements (called *yaw*).

Vertical takeoff aircraft This term refers to helicopters and tilt-rotor aircraft.

Very light jets (VLJ) Very light jets (sometimes called microjets) are aircraft that use low-noise turbofan jet engines.

Volunteer Any individual accepted to perform services by an agency that has authority to accept volunteer services when the individual performs services without promise, expectation, or receipt of compensation for services performed.

Weapons of mass destruction (WMD) (1) Any incendiary, poison gas, bomb, grenade, rocket having a propellant charge of more than 4 ounces or missile having an explosive or incendiary charge of more than 0.25 ounce, mine, or similar device; (2) any weapon that is designed or intended to cause death or serious bodily injury through the release, dissemination, or impact of toxic or poisonous chemicals or their precursors; (3) any weapon involving a disease organism; or (4) any weapon that is designed to release radiation or radioactivity at a level dangerous to human life.

Wings The main airfoils of conventional aircraft that provide lift.

Wing root The point at which an aircraft wing is joined to the fuselage.

Wing strut This structure looks like a rod or pole and is connected to the airplane between the bottom of the wing and the bottom-side area of the fuselage. The strut provides additional strength for the aircraft.

REGULATIONS TO KNOW

NFPA 402 Guide for Aircraft Rescue and Firefighting Operations

NFPA 403 Standard for Aircraft Rescue and Firefighting at Airports

NFPA 405 Recommended Practice for the Recurring Proficiency Training of Aircraft Rescue and Firefighting Services

NFPA 418 Standard for Heliports

NFPA 422 Guide for Aircraft Accident Responses

NFPA 424 Guide for Airport/Community Emergency Planning

NFPA 472 Hazardous Materials

NFPA 1001 Fire Fighter Qualifications

NFPA 1002 Driver Operator Qualifications

NFPA 1003 Airport Fire Fighter Qualifications

NFPA 1021 Fire Officer Qualifications

NFPA 1031 Fire Inspector Qualifications

NFPA 1403 Inside Live Fire Training

NFPA 1406 Outside Live Fire Training

NFPA 1500 Fire Department Safety & Health

NFPA 1521 Safety Officer Qualifications

NFPA 1561 Incident Management System

OSHA 29CFR 1910.134 Respiratory Protection

OSHA 29CFR 1910.146 Confined Space

OSHA 29CFR 1910.1030 Blood-borne Pathogens

Federal Aviation Administration in *FAR Part 139* and the *ICAO Airport Services Manual* outlines specific knowledge and skills related to the aviation firefighting environment.

ACRONYMS

AFFF Aqueous Film Forming Foam
APU Auxiliary Power Unit
ARC American Red Cross
ARFF Aircraft Rescue Firefighting
ATF Bureau of Alcohol, Tobacco and Firearms
BIA Bureau of Indian Affairs
BLEVE Boiling Liquid Expanding Vapor Explosion
BLM Bureau of Land Management
BSI Body Substance Isolation
CAP Civil Air Patrol
CDC Centers for Disease Control and Prevention
CFR Code of Federal Regulations
CIS Crisis Intervention Support
CVR Cockpit Voice Recorder
DATCP Department of Agriculture, Trade and Consumer Protection
DFG Department of Fish and Game
DFO Disaster Field Office
DHFS Department of Health and Family Services
DMA Department of Military Affairs
DNR Department of Natural Resources
DO Duty Officer
DOA Department of Administration
DOC Department of Commerce
DOD Department of Defense
DOE Department of Energy
DOJ Department of Justice
DOT Department of Transportation
DPI Department of Public Instruction
DSP Division of State Patrol
DTSC Department of Toxic Substances Control
DWD Department of Work Force Development
EAD Emergency Animal Disease
EAP Employee Assistance Program
EAS Emergency Alert System
ECB Educational Communications Board
ELT Emergency Locater Transmitter
EMAC Emergency Management Assistance Compact
EOC Emergency Operating Center
EPS Emergency Police Services
EPU Emergency Power Unit
ERP Emergency Response Plan
ERT Emergency Response Team
ESF Emergency Support Function
FAA Federal Aviation Agency
FBI Federal Bureau of Investigation
FBO Fixed-Base Operator
FCC Federal Communications Commission
FDA Federal Drug Administration
FDR Flight Data Recorder
FEMA Federal Emergency Management Agency
GA General Aviation
GIS Geographic Information System
HE Pronounced "H-E"
HF High Frequency
HMP Hazard Mitigation Plan
HMT Hazard Mitigation Team
HP Highway Patrol
IAP Individual Agency Plans
IATA International Air Transport Association

ICAO	International Civil Aviation Organization
ICP	Incident Command Post
ICS	Incident Command System
IMS	Incident Management System
JFO	Joint Field Office
JOC	Joint Operation Center
JPIC	Joint Public Information Center
JTTF	Joint Terrorism Task Force
LERP	Local Emergency Response Plan
LOS	Line of Succession
LOX	Liquid Oxygen
LZ	Landing Zone
MACS	Multi-Agency Coordination System
MCC	Mobile Command Center
MCI	Mass Casualty Incident
MOU	Memorandum of Understanding
NAWAS	National Warning System
NDA	National Defense Area
NDMS	National Disaster Medical System
NFPA	National Fire Protection Association
NGO	Nongovernmental Organization
NIMS	National Incident Management System
NIOSH	National Institute of Safety and Health
NOAA	National Oceanic and Atmospheric Administration
NPS	National Park Service
NRC	National Response Center
NRP	National Response Plan
NTSB	National Transportation Safety Board
OES	Office of Emergency Services
OIC	Officer in Charge
OSHA	The Occupational Safety and Health Act
PD	Police Department
PDA	Preliminary Damage Assessment
PIO	Public Information Officer
POWTS	Private Onsite Waste Treatment Systems
PPE	Personal Protective Equipment
PSC	Public Service Commission
PTSD	Post-Traumatic Stress Disorder
RACES	Radio Amateur Civil Emergency Services
RRC	Regional Response Center
RRP	Regional Response Plan
SAR	Search and Rescue
SEMS	Standardized Emergency Management System
SEOC	State Emergency Operations Center
SHMT	State Hazard Mitigation Team/Task Force
SO	Sheriff's Office
SOB	Souls on Board
SOG	Standard Operating Guideline
SOP	Standing Operating Procedure
STOL	Short Take Off and Landing
TIME	Transaction Information for Management of Enforcement (law enforcement teletype system)
UDSR	Uniform Disaster Situation Report
UHF	Ultra-High Frequency
USAF	United States Air Force
USCG	United States Coast Guard
USDA	United States Department of Agriculture
USFA	United States Fire Administration
USFS	United States Forest Service
USMC	United States Marine Corps
USN	United States Navy
VHF	Very-High Frequency
VLJ	Very Light Jets
VMAT	Veterinary Medical Assistance Team
VOAD	Volunteer Organizations Active in Disasters
WMD	Weapons of Mass Destruction

ADDITIONAL RESOURCES

This section contains a partial listing of information resources from which you can gather additional information related to aircraft safety, fire suppression, and various other aircraft topics.

This is only a partial listing of Web sites for aircraft safety and rescue information:

Aircraft Rescue Firefighting Working Group (ARFFWG): http://www.arffwg.org. Having members from all over the world, this nonprofit international organization dedicated to the sharing of aircraft rescue and firefighting (ARFF) information between airport firefighters, municipal fire departments, and all others concerned with aircraft firefighting.

Aircraft rescue and firefighting in Italy: http://digilander.libero.it/aircraftfire/800/homef.htm

Air Safety.Com: http://airlinesafety.com/

Aviation Fire Journal: www.aviationfirejournal.com

Boeing Fire Department, Rescue Charts and Diagrams: http://www.boeing.com/commercial/airports/rescue_fire.htm

Canadian Airport Fire Protection: http://www.cafp.net/

Emergency Disaster Management Inc.: http://www.emergencymanagement.net/

Fire Protection Association Australia: http://www.fpaa.com.au/

International Aviation Fire Protection Association: http://www.iafpa.org.uk

International Association of Fire Fighters (IAFF): http://www.iaff.org

National Fire Protection Association: http://www.nfpa.org

Polish Rescue Site : http://t.nycz.bytom.pl/

Royal Air Force Fire Service: http://www.raffireservice.co.uk/

Transport Canada: http://www.tc.gc.ca/nov1_96/index_e.htm

An excellent source of military and large-frame commercial aircraft rescue information and procedures is *Technical Order (to) Aerospace Emergency Rescue and Mishap Response Information (Emergency Services),* USAF To 00105e9. Currently, distribution of this manual is restricted and requires permission to access it. This requires filing an official registration form to open an account with the U.S. Air Force. The manual may not be reproduced, posted, or distributed without permission from the Air Force. All requests of this nature must be sent to the following e-mail address: hqafcesa/cexf@tyndall.af.mil.

Keep in mind that Web addresses change all the time. If you encounter this problem, simply search the Internet to find the updated Web address for the site you are seeking by typing "aircraft fire and rescue information" into your search engine. This should connect you to the desired sites. Many sites, such as the one maintained by the Aircraft Rescue Firefighting Working Group, provide hyperlinks to additional aircraft rescue and firefighting sites, as well as other aviation safety sites.

INDEX

Note: page numbers in **bold** indicate an illustration or table.

D

E

F

N

O

P

R

S